If the truth were a mere mathema[tical] [formula, in some] sense it would impose itself by its own power. But if Truth is Love, it calls for faith, for the 'yes' of our hearts.

POPE BENEDICT XVI

THE SHROUD OF TURIN

A CRITICAL SUMMARY OF OBSERVATIONS, DATA, AND HYPOTHESES

JOHN JACKSON, PH.D.
AND THE TURIN SHROUD CENTER OF COLORADO
www.shroudofturin.com

Project Staff

Project Director
John Jackson, PhD physics

Editor
Robert Siefker

Cover Design
Anthony Ward Arts

Writers
John Jackson PhD physics
Robert Siefker
Keith Propp, PhD physics
Rebecca Jackson
Ares Koumis
Jim Bertrand

Copy Editor / Proofreaders
Mary Snapp
Mary Ann Siefker

Special Acknowledgments

We wish to gratefully acknowledge the contributions of several individuals and organizations. First, we would like to acknowledge Dan Spicer, PhD in physics, and David Fornof for contributing to the construction of Version 1 of the *Critical Summary*, originally published on the TSC website. We express our gratitude to Shroud Historian Jack Markwardt for reviewing the History Section and providing valuable comments. We thank Shroud scientist Giulio Fanti for furnishing important images and helpful comments. We most especially thank Barrie Schwortz and the STERA organization for providing high resolution Shroud and **Shroud of Turin Research Project (STURP)** historical photographs. Barrie was a colleague of John Jackson on the STURP project and today administers the important Shroud research repository site http://www.shroud.com/. Additional acknowledgments are included in the Photograph and Image Credits Section.

Edition 1

Edition 1 corresponds to online Version 4.0 of the Critical Summary published on the TSC website.

First printing in paperback 2017

Copyright ©2017 The Turin Shroud Center of Colorado (TSC)

ISBN 978-0-692-88573-4

All rights reserved. No part of this publication may be reproduced, stored in a retrieval system, or transmitted in any form by any means, electronic, mechanical, photocopying or otherwise, without first obtaining written permission of the copyright owner.

Printed in the United States of America

Scripture texts in this work are taken from the *New American Bible,* revised edition ©2010, 1991, 1986, 1979 Confraternity of Christian Doctrine, Washington, D.C. and are used by permission of the copyright owner. All Rights Reserved. No part of the New American Bible may be reproduced in any form without the permission in writing of the copyright owner.

All Shroud and STURP photographs are ©1978 STERA, Inc., unless otherwise noted (see photograph and image credits.

The Turin Shroud Center of Colorado

www.shroudofturin.com

The Turin Shroud Center of Colorado is a proud education partner of the *American Confraternity of the Holy Shroud* (ACHS).

www.shroudconfraternity.org

Table of Contents

Introduction		3
Section 1: Historical Evidence	(Items H1-H28)	7
Section 2: Medical Forensic Evidence	(Items M1-M17)	45
Section 3: Linen Cloth Evidence	(Items L1-L14)	55
Section 4: Image Characteristic Evidence	(Items C1-C6 / B1-B11)	67
Section 5: Image-Formation Hypotheses	(F1-F10)	79
Section 6: Rating the Image-Formation Hypotheses		87
Section 7: Dating the Shroud		89
Concluding Comments		97
Appendix 1: STURP Team Members		99
Appendix 2: Rating Details for Image-Formation Hypotheses		101
Appendix 3: Evidence Revision Log		112
References		115
Photograph and Image Credits		131

(Fig. 1) Christ Pantocrator icon, dated to ca. 550, from St. Catherine's Monastery in the Sinai Desert of Egypt. See Item H9 in the *History Section* for an extended discussion.

Introduction

In 1978 PhD in physics John Jackson, who today serves as the president of the **Turin Shroud Center of Colorado**, led a large research team from the United States on an historic project to study the Shroud in Turin, Italy. The team, under the auspices of the **Shroud of Turin Research Project (STURP)**, was given unprecedented hands-on access to the Shroud. For 120 continuous hours, the Shroud was examined in depth. Such direct research access to the Shroud had not been given prior to this, nor has it been given since. The STURP team consisted of outstanding scientists, research assistants and professional photographers. Appendix 1 lists the names and home organizations of the STURP team members. The team used advanced scientific instruments for their five days and nights of examining the Shroud. Among the methods used to gather data were direct microscopy, infrared spectrometry, X-ray fluorescence spectrometry, X-ray radiography, thermography, and ultraviolet fluorescence spectrometry. In addition, a broad spectrum of photographic data was collected. Ultraviolet fluorescence photographs, raking-light photographs, normal front-lit photographs and backlit photographs of the entire Shroud were taken. The STURP team also collected sticky tape samples from the surface of the Shroud cloth as well as thread samples. The Minnesota Mining and Manufacturing Corporation (3M) designed and produced a special tape specifically for the STURP project: an amorphous, inert, pure-hydrocarbon adhesive that would not contaminate the Shroud samples. All of these samples were retained and returned to the United States for further studies. The subsequent studies of the tape samples were carried out using microscopy, pyrolysis-mass-spectrometry, laser-microbe Raman analysis and various methods of microchemical testing. The results of the STURP research were published in twenty peer-reviewed scientific journal articles over the course of four years following the team's work in Turin.

The purpose of the **Critical Summary** is to provide an up-to-date summary of what is known about the Shroud. The Critical Summary includes many pieces of data that trace their source back to the research conducted by the STURP team, as well as data from the broad spectrum of other scientific and historical research that has been conducted on the Shroud, both before and after the STURP expedition to Turin. One thing is certain: The Shroud is an artifact that plainly exists. It is of interest because it is the informed judgment of many that the Shroud is the actual burial cloth of Jesus of Nazareth.

This is a remarkable claim. Can it be true? Only through a serious examination of the evidence can the inquirer begin to come to a justified judgment regarding the answer to that potentially life-changing question. The Critical Summary is designed to help the inquirer begin that examination.

How the Critical Summary is organized

The Critical Summary is designed to be maintained and evolve over time as important new historic, scientific and forensic evidence on the Shroud is discovered. Because of this we would characterize the Critical Summary as a "living" summary of the high points of Shroud research. Each new edition will be given a new version number. New editions corresponding to the inclusion of extremely important new research findings regarding the Shroud will be given a cardinal number designation, for example **Version 4.0**. Less important new evidence will be documented in an on-line only edition (i.e. **Version 4.1**), that may be accessed on the **TSC** website (http://www.shroudofturin.com). All changes to Shroud evidence documented in the Critical Summary can be tracked through reference to the **Evidence Revision Log** contained in Appendix 3.

Data items presented in Sections 2, 3 and 4 are based on forensic and scientific study of the Shroud. The scientific method requires that such evidence be presented as a proposition that may, by its very nature, be subject to further study and possible revision. At a minimum, all such items of evidence must be evaluated and assessed. The Critical Summary uses the following ratings for each item of forensic evidence in these three sections:

Class 1: This rating is given to items of evidence that are firmly supported by scientific and/or forensic research.

Class 2: This rating is given to an item of evidence generally supported by scientific and/or forensic research, but the item requires additional confirming research in order to be upgraded to Class 1.

Class 3: This rating is given where the item is documented or reported as Shroud evidence but remains disputed by many researchers including the TSC organization.

INTRODUCTION

Evidence Presentation and Photographs

Shown below is an example of how an item of evidence is presented and how the item Identification is linked to the reference section of the Critical Summary. The example shown is from **Section 3: Linen Cloth Evidence**. Following the example of how the evidence is presented are two photographs that show how Shroud photographs are presented in the Critical Summary. The first photograph shows the image side of the Shroud, as it would be viewed in natural light. This photograph shows the locations where the frontal and dorsal images of a human body are imprinted on the cloth. Noted on this photograph are also the prominent burn and scorch lines left on the cloth by a fire that damaged the Shroud in the year 1532. The second photograph is a negative photograph of the Shroud showing the upper body portion of the Shroud **Frontal Image**. As you read through the Critical Summary it should be kept in mind that the imprinted frontal and dorsal images on the cloth are a mirror image of any actual body that might have been wrapped in the Shroud. In many cases the photographs in the Critical Summary have been reversed so that images are presented as if an actual body is being viewed. To help the reader avoid confusion key Shroud photographs have been labeled to indicate whether the **Shroud Image (SI)** or **Body Image (BI)** view is being presented.

(SI) = Shroud Image

(BI) = Body Image

The following three benchmarks will also be helpful in establishing orientation when viewing frontal negative photographs of **Body Images (BI)**.

1. The blood flow on the forehead when looking at the body resembles the letter "3".
2. The chest wound is on the right side of the body.
3. The **left hand** of the body crosses the **right hand**.

It will also be helpful to remember that in positive photographs of the actual Shroud, blood and blood flows are **dark** in color. In black and white photographic negatives of the Shroud blood appears to be **white**.

Example of How a Data Item is Presented

Item and Class Rating
↓

The Evidence is presented in bold starting at the left margin. Elaborating comments are given below the Evidence and are indented.
↓

ID R	Evidence/Comment
L1 1	**The Shroud conservation project of 2002 stabilized the layout of the Shroud by stretching it out for flat storage. The reported post-preservation dimensions are 14' 6" x 3' 9' (4.42 x 1.14 m).**
	The Shroud was not likely woven to these particular specifications. A more likely weaving specification for the Shroud of **8 cubits long by 2 cubits wide** would conform closely with the ancient Assyrian cubit of approximately 21.7 inches (55.1 cm) that was used in the area of Palestine in the first century [1].

References are listed using the same item identification number as that used for the evidence. For example, the references for this item would be listed in the References under Item L1:

Ref-L1 1. Eleanor Guralnick, "Sargonid Sculpture and the Late Assyrian Cubit," *Iraq Journal* 58 (1996): 89-103.

INTRODUCTION

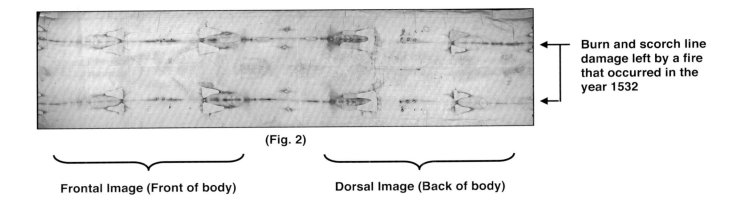

(Fig. 2) — Burn and scorch line damage left by a fire that occurred in the year 1532

Frontal Image (Front of body)   Dorsal Image (Back of body)

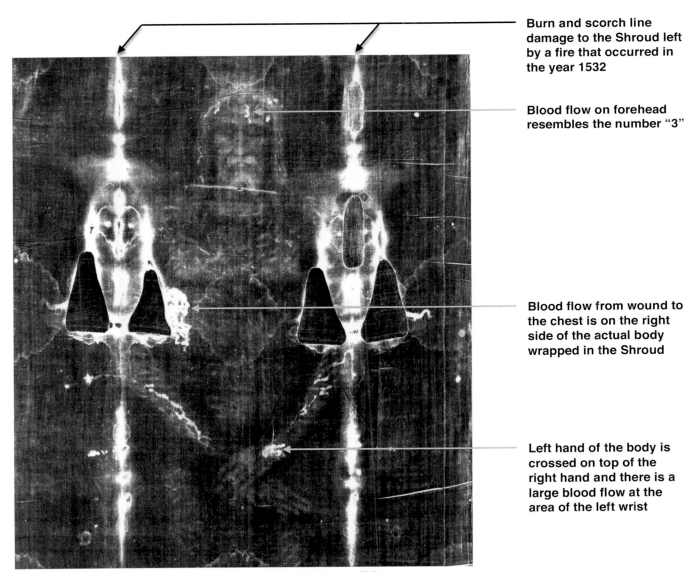

(Fig. 3) Negative of Shroud frontal image (BI)

- Burn and scorch line damage to the Shroud left by a fire that occurred in the year 1532
- Blood flow on forehead resembles the number "3"
- Blood flow from wound to the chest is on the right side of the actual body wrapped in the Shroud
- Left hand of the body is crossed on top of the right hand and there is a large blood flow at the area of the left wrist

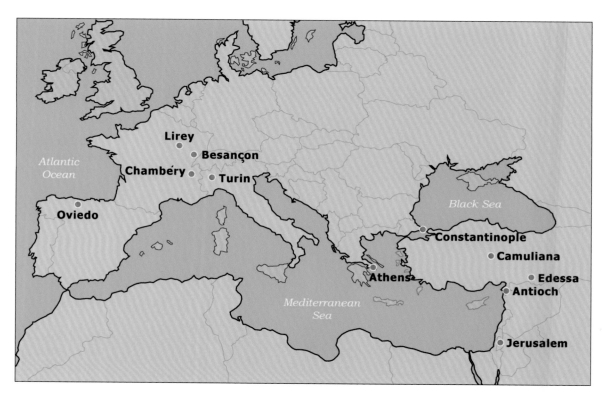

(Fig. 4) Map of key locations related to Shroud history

Section 1: Historical Evidence

The Shroud's first public exposition in Europe occurred in Lirey, France in 1355. After this date, the Shroud's history is generally well documented. Because of this, we include only major historical data points after the Shroud's first public appearance in Europe. Our effort in this section of the Critical Summary is primarily focused on historical data prior to 1355 that casts light on the earlier history of the Shroud.

Our effort in constructing this table has been guided by one unifying goal…. to present an historical "start". Our historical "start" consists of the following:

1. A presentation of the key historical data points that have been associated with the Shroud by various historians, starting with the Gospel narrative of the "burial clothes" of Jesus up to the first public exposition of the Shroud in Europe in 1355. Here we emphasize evidence that attests to the Shroud being in Constantinople in the year 1203.

2. An abbreviated presentation of key historical data points related to the Shroud in Europe subsequent to 1355. There are many detailed, and easily accessible, chronological-historical timelines of the Shroud's history in Europe, as well as numerous full-length books, and we refer you to these. We include only the key data points for the Shroud's European history with which all serious inquirers should be familiar.

3. The Shroud's history prior to 1203 is tied to historical data points that have been proposed by historians to have a relation to the Shroud, some more compelling than others. The reader is encouraged to keep in mind that "later events" often cast a great deal of light on "earlier events". We therefore recommend that the reader first study the historical evidence as presented. Then take a second look at the history of the Shroud from 1203 backwards to 33 AD. We think that examples will be seen where "later" data casts considerable light on "earlier" historical data.

ID	Narrative on the Historical Evidence
H1	**ca. 33 AD: The crucifixion, death and burial of Jesus of Nazareth** The most frequently suggested date of the death of Jesus in Jerusalem is Friday April 3, 33 AD. [1] All of the Canonical Gospels state that the dead body of Jesus was "wrapped" in a linen cloth and buried in a rock tomb. The Synoptic Gospel of Matthew reports as follows: *When it was evening, there came a rich man from Arimathea named Joseph, who was himself a disciple of Jesus. He went to Pilate and asked for the body of Jesus; then Pilate ordered it to be handed over. Taking the body, Joseph wrapped it in clean linen and laid it in his new tomb that he had hewn in the rock. Then he rolled a huge stone across the entrance to the tomb and departed.* *Matthew 27: 57-60* The Gospel of John reports on the events of Sunday morning, and again, the linen burial cloth is prominently mentioned: *On the first day of the week, Mary of Magdala came to the tomb early in the morning, while it was still dark, and saw the stone removed from the tomb. So she ran and went to Simon Peter and to the other disciple whom Jesus loved, and told them, "They have taken the Lord from the tomb, and we don't know where they put him." So Peter and the other disciple went out and came to the tomb. They both ran, but the other disciple ran faster than Peter and arrived at the tomb first, he bent down and saw the burial cloths there, but did not go in.* *John 20:1-5*

33 – 68 AD: The Shroud and the Apostolic Period up to the Death of Saint Peter

Matthew's Gospel records Jesus' final words to his apostles after the resurrection, the words of the "**Great Commission**":

> *Then Jesus approached and said to them, "All power in heaven and on earth has been given to me. Go, therefore, and make disciples of all nations, baptizing them in the name of the Father, and of the Son, and of the holy spirit, teaching them to observe all that I have commanded you. And behold, I am with you always, until the end of the age."*
>
> *Matthew 28: 18-20*

Before his crucifixion and death Jesus had spoken the following words to his apostle Simon the fisherman: **"And so I say to you, you are Peter, and upon this rock I will build my church, and the gates of the netherworld shall not prevail against it."** (Mt 16:18) "Rock" is translated Cephas in Aramaic, and Petros or Peter in Greek. Thus Peter, in his leadership role in the early church, was given special responsibility for the **Great Commission**. The Acts of the Apostles portrays Peter as fully embracing this responsibility by becoming the most active of the original apostolic missionaries, venturing widely to evangelize and preach the Gospel. Peter is known to have focused, at least in his early missionary work, on taking the Gospel to the Jews of Palestine and to the Jewish communities in the Roman Empire. [1] Saul of Tarsus, who became **Paul the Apostle,** was also a great missionary leader along with Peter. Paul became known as the "Apostle to the Gentiles."

A set of extraordinary (providential?) historical events worked to set the table for the rapid wide-ranging spread of the Christian Gospel during the Apostolic Era. On 9 August 48 BC, eighty-five (85) years before Jesus gave the Great Commission, the decisive battle of the Roman Civil war took place at Pharsalus in central Greece. In that battle, the Roman general **Julius Caesar** and his legions defeated the much larger army of the Roman Senate, commanded by his great rival, **Pompey the Great**. The defeated Pompey gathered a small remnant of his army, and in an endeavor to survive and fight another day, escaped by boat across the Mediterranean Sea to Egypt. Caesar soon pursued Pompey with a relatively small force of his own. On landing in the Egyptian city of Alexandria, Caesar learned that Pompey had already been murdered and that he, Caesar, had sailed into mortal danger. Egyptians were caught up in their own civil war, and forces engaged in that struggle now saw Caesar's presence in Egypt as a threat to their own cause. Soon a force of 20,000 Egyptian soldiers on foot and 2,000 cavalry converged on Alexandria to exterminate Caesar and his men. Caesar was outnumbered and in threat of annihilation.

Then a history-changing event occurred. From just south of the Roman Province of Judea a relatively obscure local leader known by the name **Antipater the Idumaen** came to Caesar's rescue. As recorded by the first century Romano-Jewish historian Flavius Josephus, *"**Antipater came to him, conducting three thousand of the Jews, armed men. He had also taken care the principal men of the Arabians should come to his assistance; and on his account, it was that all the Syrians assisted him also."*** [2] Antipater also rallied local members of the large Jewish population of Alexandria to support Caesar. Caesar's resulting survival, coupled with the death of Pompey, foreshadowed the end of the Roman Republic and the beginning of the Roman Empire under Caesar's adopted son Octavius who would become known as the first Roman Emperor… Caesar rewarded Antipater for coming to his rescue by naming him procurator of the Roman Province of Judea that included the city of Jerusalem. Thus, Caesar gave birth to the **Herodian Dynasty.** Antipater was the father of **Herod the Great,** and the great-grandfather of **Herod Agrippa.** Herod the Great was officially known as "The King of the Jews" decades before the birth of Jesus, who would later have that same title written on a piece of wood and nailed to his cross. Herod the Great is also known for the New Testament story in which he slaughtered every male infant in Bethlehem in an attempt to kill the infant Jesus. Herod Agrippa imprisoned Peter and planned his execution to suppress the spread of Christianity. Agrippa's uncle was **Herod Antipas,** who is known for the murder of John the Baptist and his role in the events surrounding the crucifixion of Jesus. It would appear that the legacy of Antipater was detrimental to Christianity; yet, the opposite is true. By saving Caesar with the help of the Jews, Antipater influenced Caesar to issue a series of decrees that directly supported the remarkable early growth of Christianity. Caesar's decrees extended religious freedom to the Jews in all the regions controlled by Rome. Caesar's adopted son, Octavius, known to posterity as **Caesar Augustus** (reign 27 BC – 14 AD), actively enforced the edicts of his father that protected the rights of the Jews in the empire. These edicts gave Judaism the status of a *Religio Licita* (permitted religion). [3] At a stroke, free, vibrant and broadly dispersed Jewish communities surfaced throughout the Roman Empire. These Jewish communities, their synagogues, and adjacent populations of Gentiles became prime targets in subsequent years for the Gospel-spreading work of Peter, Paul and other early Christian evangelists working to carry out the Great Commission.

(Fig. 5) Map showing the extraordinary spread of Christianity during the Apostolic Era

1. **Discipline of the Secret**: The early history of the Shroud has been obscured by an ancient church custom that seventeenth-century theologians labeled the "*Discipline of the Secret*". [4] Pursuant to this custom, early Christian leaders, when speaking of Christian tenets, doctrines, mysteries, and rites, employed coded language, symbolic representations, metaphorical expressions, and allegorical narratives so as to make their words understandable only to advanced believers. This custom was adopted in compliance with Christ's commandment ***"Do not give what is holy to dogs, or throw your pearls before swine, lest they trample them underfoot, and turn and tear you to pieces."*** (Mt. 7:6) For example, the early Church referred to Jesus as a "fish", the Eucharist as "the honey sweet food of the redeemer", the consecrated bread and wine as "the symbols", and baptism as the "seal". Practice of the *Discipline* is evidenced in the ecclesiastical writings of the early church fathers. Clement of Alexandria wrote "**the mysteries are delivered mystically, that what is spoken may be in the mouth of the speaker; rather not in his voice, but in his understanding**." [5][6] The *Discipline of the Secret* continued to be practiced until the Church became fully established in the fifth century. By applying the precepts of this Church custom, historian Jack Markwardt has identified Christian references to Jesus' image-bearing burial shroud and its employment in early evangelical missionary work.

2. **Antioch in Syria:** The death of Stephen, venerated as the first martyr of Christianity, by stoning is generally held to have taken place in the year 34 in Jerusalem. His death marked a time of persecution for early Christians. As a result, some of the followers of Jesus fled from the city and traveled as far as Phoenicia, Cyprus, and Antioch, preaching to the Jews in those regions. [7][8] Barnabas apparently was also sent to Antioch, likely by Peter.

    - At the time, Antioch was the third most important city in the Roman Empire after the city of Rome and the Egyptian city of Alexandria. [9]

    - In approximately the year 40, under the leadership of Barnabas and Paul, a number of Christian missionaries shifted their focus to the Gentiles. Antioch became the base for those great missionaries. [10]

    - After 70 AD Antioch hosted the world's largest Christian community, and the term "Christian" was coined there. [11] Antioch was to be called the "Cradle of Christianity."

    - St. Luke was a native of Antioch and wrote his Gospel and the *Acts of the Apostles* in that city.

3. **ca. 40 The Gospel of the Hebrews and Peter:** The Shroud scholar Maurus Green, OSB (b 1929- d 2001) wrote: "*The fact that our Lord's burial cloths and their arrangement were the first material evidence of the Resurrection would point to their preservation despite their defiling nature – anything to do with a corpse being impure to the Jews.*" [12] St. Jerome is the first to explicitly record this preservation. He quotes the lost Gospel of the Hebrews that may have predated the Gospel of Matthew, to the effect that the Lord confided the Shroud (sindon) "**to the servant of the priest**." [13] [14] [15] Some have interpreted this passage from the Gospel of the Hebrews to mean the Shroud was given to Peter, or at least ended up in his possession. Maybe there is a further hint supporting this interpretation associated with a reading closely associated with the Christian Feast of Corpus Christi that states that Jesus came as "...**a high priest of the good things that have come.**" (Hebrews 9:11) Peter was the "rock", and thus, can be viewed as the "chief servant" of Jesus. There is a hint here, and its oblique nature is in keeping with the *Discipline of the Secret.* [16]

4. **The Missionary Work of Peter:** Unlike Paul, who engaged in three specific and datable missionary journeys, Peter's specific missionary activities and dates remain somewhat vague. The first epistle of Peter, however, provides clues regarding Peter's missionary travels. This letter from Rome, if written directly by Peter, would be dated 60-63. More likely the letter was written between 70-90 by a disciple carrying on the heritage of Peter in Rome. The letter is regarded as one of the New Testament's most beautiful and compelling books: "Its profound Christology, vision of the church and ardent instruction on Christian life in the world richly express the meaning of the Gospel." [17] The epistle's opening address indicates where Peter had likely traveled:

> "**Peter, an apostle of Jesus Christ, to the chosen sojourners of the dispersion in Pontus, Galatia, Cappadocia, Asia, and Bithynia, in the foreknowledge of God the Father, through sanctification by the Spirit, for obedience and sprinkling with the blood of Jesus Christ: may grace and peace be yours in abundance.**"
>
> *1 Pet 1:1-2*

Additionally, the *Doctrine of the Apostles*, a Syriac work that is datable to the fifth or sixth century, recites that: "*Antioch, and Syria, and Cilicia, and Galatia, even to Pontus, received the Apostles' Hand of Priesthood from Simon Cephas, who himself laid the foundation of the church there, and was Priest, and ministered there up to the time when he went up from thence to Rome.*" [18] The following are traditionally accepted dates for Peter's travels and whereabouts: [19][20]

- It would appear that Peter first went to Antioch in ca. 35 when he supported the founding of the church there. Bartholomew and Paul arrived in Antioch later. It is likely that Peter returned to Jerusalem in ca. 40, and he was certainly in that city in 42.

- In the year 42 great dangers threatened Peter. Herod Agrippa, the great-grandson of Antipater, who had been instrumental in opening the door of the Roman Empire for Christian evangelization, acted to suppress evangelization of the Gospel. First Herod killed the apostle James, the brother of John; then, he turned on Peter and had him arrested and threatened him with the same fate. The *Acts of the Apostles* reports that Peter miraculously escaped and secretly left Jerusalem to travel to "**another place**." (Acts 12:17) Herod, subsequently, put to death the guards who "allowed" Peter to escape. Agrippa was deadly earnest, and Peter was in deadly peril. The *Acts* state that Peter went to "Caesarea and stayed there." Many biblical scholars doubt that he stayed in Caesarea for long. To escape to a neighboring province would have been to invite extradition. It has been hypothesized that the most likely place in the world to harbor an escaped Jewish prisoner was also the home of a vast Jewish population and a fertile field for evangelization – Rome. [21]

- Herod Agrippa died in the year 44, and Peter is recorded to be back in Jerusalem to attend the Jerusalem Council ca. 49-50. At the Council of Jerusalem, it was agreed that Gentiles could be accepted as Christians without full adherence to the Mosaic Laws, particularly circumcision.

- After the Council of Jerusalem tradition has Peter back in Antioch until ca. 54-55 when he was to leave Antioch for his second Journey to Rome.

- After 55 tradition holds that Peter evangelized in Rome and Italy until his crucifixion at the hands of the Roman Emperor Nero in ca. 64 – 68.

	5. **Paul's Letter to the Galatians:** Paul's letter to the Galatians was most likely written while he was in Ephesus ca. 54-55. [22] He wrote: ***"O stupid Galatians! Who has bewitched you, before whose eyes Jesus Christ was publicly portrayed as crucified?*** (Gal 3:1) In his letter Paul exhorts the Galatians to hold tight to the core of the Gospel that had been preached to them by someone other than himself who had even "exhibited" the crucified Christ to them. Many Shroud scholars think Paul is referring to an earlier time when "Peter used the Shroud during his missionary work in the region of Galatia." [23]
H3	**68 – 70 AD: "Image of our Holy Lord and Savior" and other Church objects removed from Jerusalem**
	If Peter had custody of the Shroud, he would not have exposed it to the dangers of a Nero-ruled Rome when he left for his second journey to that city in ca. 55. If he had the Shroud in his possession in Antioch, he may have simply left it with the church of Antioch. However, the church of Jerusalem had been at peace since the death of King Herod Agrippa in 44, and Peter may have entrusted the Shroud to Jesus' relative, James the Just, the leader of the Jerusalem church.

In 62, James was killed, and four years later open hostilities broke out around Jerusalem between Jewish zealots and their Roman rulers. [1] Seeing the deteriorating situation, many in the Church of Jerusalem fled the city, some undoubtedly for Antioch or other locations in Syria. [2] Such flight is recounted in the **Sermon of Athanasius**, a text ascribed to Saint Athanasius, the Bishop of Alexandria (ca. 328 -373), and read to the Second Council of Nicaea in 787. Athanasius' sermon reflects a tradition holding that in the year 68 an "*image of our Lord and Savior at full length*" [3] was taken from Jerusalem and moved to Syria:

> *"But two years before Titus and Vespasian sacked the city, the faithful and disciples of Christ were warned by the Holy Spirit to depart from the city and go to the kingdom of King Agrippa II, because at that time Agrippa II was a Roman ally. Leaving the city, they went to his regions and carried everything relating to our faith. At that time even the icon with certain other ecclesiastical objects were moved and they today still remain in Syria. I possess this information as handed down to me from my migrating parents and by hereditary right. It is plain and certain why the icon of our Holy Lord and Savior came from Judaea to Syria.* [4] |
H4	**70 AD: Siege of Jerusalem**
	In 70 A.D., the final Roman siege of Jerusalem was led by the future Emperor Titus. The city was sacked, and the Jewish Second Temple was destroyed along with most of the rest of the city. The Siege of Jerusalem was aimed at Jewish zealots who occupied major parts of the city, but it also caught up any Christians that might have still remained there. Antioch would have been a prime destination for Christian refugees fleeing from the areas of besieged Jerusalem. [1]
H5	**70 – 306 AD: Pre-Constantine Era**
	Commencing with the attacks by Nero in 64, the executioner of Peter and Paul, and continuing for the next two and a half centuries, Roman rulers either instituted persecutions or tolerated local persecution against Christianity. Prior to 180 the foremost persecutors of the church, besides Nero, were the emperors Domitian (81-96), Trajan (98-117), Hadrian (117-138) and Marcus Aurelius (161-180).

1. **100-115 Persecution in Antioch:** Early persecutions fell particularly hard upon the church of Antioch. The city produced several martyrs, including its bishop Ignatius, who at the direction of Trajan was transported to Rome where he was killed in the arena by wild beasts. [1]

2. **135 Jerusalem transformed into a pagan city:** In AD 135 the Roman emperor Hadrian re-founded Jerusalem as a pagan Roman city named **Aelia Capitolina.** [2] Sites considered holy by the Christians, as well as Jews, were systematically targeted and desecrated by the construction of pagan structures. At Hadrian's instructions, a pagan temple was built over the place of the tomb of Jesus. [3] Eusebius of Caesarea (AD 260-340), an early church historian, recalled that he had seen with his own eyes how the pagan temple had been built over Jesus' tomb *"above the ground they constructed what could be described with terrible truth as a tomb for souls, building a gloomy alcove for dead idols in honor of the licentious demon Aphrodite (Venus), and then pouring cursed libations there over impious and profane altars."* Later, Saint Jerome (AD 347-420) stated that even the crucifixion site of Jesus was marked by a pagan shrine*: "On the rock of the cross a statue of Venus made of marble was venerated by the pagans."* |

3. **ca. 190 The conversion of Abgar VIII (the Great) of Edessa:** In 180 Commodus, the son of Marcus Aurelius, became emperor. Married to a Christian and devoted entirely to the pursuit of personal pleasures, Commodus provided thirteen years of profound peace for the Church. In 177 Abgar VIII, known as **Abgar the Great**, became king of the Mesopotamian kingdom of Oshroene and ruled from its capital city Edessa, which was approximately 145 miles northeast of Antioch. During his reign, which extended to 212, a Christian church was built in Edessa, Christian imagery appeared on Royal coinage, and a Christian synod was reportedly held in his kingdom. [4] The conversion of King Abgar VIII (the Great) was reported in both the *Liber Potificalis* and the writings of the Venerable Bede, both of which describe how a *"British King Lucius"* sent a letter to Pope Eleutherius (pontificate: ca. 174 to 189) in which the king asked to be baptized and made a Christian. However, late second-century Britain was not ruled by a king, but by Imperial Rome.

(Fig. 6) Inscription of Abercius

Adolph Harnack, a respected biblical scholar, deduced that the reference to a *"Britannio Rege Lucio"* was in fact, an allusion to the *"Britio Edessenorum"*, Edessa's citadel, and to King Lucius Ælius Septimuius Megas Abgarus VIII, that is to King Abgar the Great of Edessa. Thus, it appears that a papal mission brought Christianity to the city of Edessa at the specific invitation of its king. This mission had to be completed prior to Commodus' death in late 192, as six months later the new emperor was the anti-Christian Septimius Severus.

The only documented late second-century ecclesiastical journey, which began in Rome and ended in Mesopotamia, was that made by Avircius Marcellus, the Bishop of Phrygian Hieropolis. Some writings estimate Avircius' death as occurring ca. 167, but they provide no reason for this dating, and inconsistently, add that he was the author of a treatise on Montanism (a second century heretical sect) which is datable to about 193. The best interpretation of the corpus of historical sources rendered by scholars William Ramsey, J. Tixeront, and Johannes Quasten is that Avircius did not die until the final decade of the second century or, perhaps, even the first decade of the third century. [5] The later date is significant because the famous monumental inscription known as the *Inscription of Abercius*, datable to 192, is attributed to the Bishop Avircius Marcellus of Phrygian Hieropolis (the picture is of a cast of the inscription from the Lateran Museum Collection, Rome).

This monument records a metaphorical summary of his travels. Scholars widely agree with the esteemed theologian Johannes Quasten that the *Inscription of Abercius* was **"written in a mystical and symbolic style, according to the Discipline of the Secret, to conceal its Christian character from the uninitiated."** [6] The inscription relates that the pope summoned the author to Rome *"to see a Queen Golden-robed and Golden-sandaled"*. It has been hypothesized that this queen was Abgar VIII's wife Shalmath, and that it was she who carried the letter to Rome in which the king requested baptism from the pope. The *Inscription of Abercius* then relates that the author traveled to Mesopotamia and that he saw all of the cities of Syria, which would have included Antioch and Edessa, which was then considered the chief city of eastern Syria. Further, the *Inscription of Abercius* recounts that the author traveled with someone named "Paul," likely the cleric "Palut", who would ultimately become the first bishop of Edessa. Most importantly, the *Inscription* discloses that at some point on the journey the author was provided with **"a fish of exceeding great size"** which possessed **"wine of great virtue"** that **"was mingled with bread."** It should be recalled that by this time Christians had begun to use the sign of a fish as a symbol of Christ to mark meeting places, tombs and as code to distinguish friend from foe. The Greek word for fish is ΙΧΘΥΣ, which is the acronym in Greek for **Jesus, Christ, Son of God, Savior.** So, in using the code language of a "fish", the *Inscription of Abercius* metaphorically reveals that the author might have been in possession of a sizable image of Jesus, which presented not only his body image ("bread") but also his bloodstains ("wine"). It is the hypothesis of Shroud historian Jack Markwardt that the author of the *Inscription of Abercius* took the Shroud from Antioch to Edessa to support the conversion of King Abgar VIII (the Great), and that the Shroud was subsequently returned to Antioch. [7]

4. **ca. 220 The Hymn of the Pearl:** Intriguing support for the hypothesis that the Shroud was used in Edessa to support the conversion of Abgar the Great is provided by the *Hymn of the Pearl,* a poem datable to the

first half of the third century. [8] Written no later than 224, the hymn, like the *Inscription of Abercius*, is fashioned in accordance with the *Discipline of the Secret,* and its promotion of certain heretical tenets strongly suggests it to be the work of Bardaisan, a Gnostic Christian philosopher who was born in Edessa and reportedly attended school with the future Abgar the Great. As the king's life-long friend and a frequent visitor to the Royal Court, Bardaisan would have likely viewed the Shroud if Avircius Marcellus had brought it to Edessa. The poem's mystical text presents a perfect Christian allegory: The protagonist prince, the son of a king, represents Jesus, the Son of God, and his robe represents the Shroud. When the poem begins, the robe is imageless. The mission assigned to the prince by his father, wrestling a pearl from the hold of a serpent, represents the mission entrusted to Christ by his heavenly Father – the redemption of humanity from the hold of the serpentine Satan. It is only after the prince's mission has been successfully completed that he is able to see that his robe now presents an image of himself: [9]

> ***On a sudden, as I faced it,***
> ***The garment seemed to me like a mirror of myself.***
> ***I saw it all in my whole self,***
> ***Moreover I faced my whole self in it,***
> ***For we were two in distinction***
> ***And yet again one in one likeness.*** [10]
>
> ***And the image of the King of kings***
> ***Was depicted in full all over it.*** [11]

The prince's declaration that the robe displays his "whole self" appears to be consistent with the Shroud and *Inscription of Abercius*' description of a fish of "exceeding great size." The hymn's reference to the robe's image as being that of the "King of Kings" reflects the passage in the Book of Revelation where Jesus is named the King of Kings: **They will fight with the Lamb, but the Lamb will conquer them, for he is Lord of lords and King of kings, and those with him are called, chosen, and faithful. (Rev 17:14)** [12] Indeed there appear to be multiple hints from Edessa that the Shroud visited the city at the time of Abgar the Great to support the king's conversion. [13]

5. **ca. 190-306 Continued Persecutions and the rise of Arianism:** With the Shroud most likely back in Antioch, Roman persecutions of Christianity were renewed and continued for more than a century. These attacks became most intense during the reigns of the emperors Septimius Severus (193-211), Maximinus the Thracian (235-238), Decius (249-251), Valerian (253-260) and Diocletian (284-305). Diocletian's was the last and most furious of the ten waves of Christian persecution during the pre-Constantine era. He issued a decree in the year 303 for all church buildings to be destroyed and for all copies of the Christian Bible to be seized and burned. Under Diocletian, Christians were systematically deprived of civil rights and denied any form of government employment. In the middle of the third century, the Bishop of Antioch was arrested and died in prison, and even after Constantine had taken control of the Western Empire in 306, Christians in Antioch continued to be persecuted by his co-emperor Galerius (305-311). [14] Throughout these persecutions, Christian leaders were compelled to scrupulously observe and enforce the *Discipline of the Secret* and keep concealed the existence and whereabouts of any surviving burial linens of Christ. Even religious believers who held that images of God or Christ were prohibited as being sacrilegious (early iconoclasts) were a risk to any image-bearing relic.

In ca. 260, Paul of Samosata became Bishop of Antioch and began to advocate the doctrine of a non-Trinitarian God. Although he was relieved of his ecclesiastical duties, a preacher Lucian of Antioch, began to teach that Jesus, as the Son of the Father, could not have existed for all eternity. This doctrine advanced by Lucian, the student of the deacon Arius, became known as **Arianism**, and the heretical doctrine was soon embraced by a majority of Antiochene Christians. [15]

| H6 | **306 – 361 AD: Constantine Era** |

The Roman Emperor Constantine (the Great) reigned from AD 306 to 337. Constantine was the first Roman Emperor to convert to Christianity. In the year 313, he issued the **Edict of Milan** that made the open practice of Christianity legal in the territories of the Roman Empire and nominally ended Roman persecution of Christianity. In the year 324 Constantine moved the capital of the empire from Rome to the eastern city originally known as Byzantium and renamed the city **Constantinople**. The eastern part of the Roman Empire was to become known as the **Byzantine Empire** after the fall of the western part of the Empire in 476. The Byzantine Empire would survive for over 10 centuries, with Constantinople as its capital until its fall on May 29, 1453 to an invading army of the Muslim Ottoman Empire. In 1930, a law was enacted in Turkey that renamed the City of Constantinople to Istanbul.

1. **ca. 324:** Constantine organized a regional synod of Orthodox Bishops that elected one of their own as Antioch's Bishop and condemned Arius. At roughly the same time Constantine ordered that a church replace the pagan temple built over Jesus' tomb in Jerusalem. This church was to become known as the Church of the Holy Sepulchre. Constantine's mother, Helen, also a convert to Christianity, went to Jerusalem during the construction of the new church in search of Passion Relics. What she found she appropriated in the name of the Empire. It was claimed she found three nails from the Crucifixion and other relics, including remnants of "the true cross", in a Christian Shrine near the Holy Tomb of Christ. She sent two of the Holy Nails to her son Constantine along with a remnant of the "true cross". One nail ended up attached to his battle helmet, another was used to fashion a bridle for his horse, and the piece of the "true cross" was incorporated into a statue he had constructed of himself. [1] Such Imperial appropriation of relics would put on notice anyone in local churches that had custody of any church "pearls" and strongly reinforce the *Discipline of the Secret*. Also, the doctrine of iconoclasm was in play. This doctrine originated in Judaism and held that religious images might constitute idolatry or encourage profane forms of worship. During Constantine's era, even before the more institutional forms of iconoclasm that would arise in later centuries in the Byzantine Empire, some Christians adhered to the iconoclast doctrine, especially in the Eastern Church. This too would reinforce the *Discipline of the Secret* with respect to any image-bearing relic of Christ. [2][3]

2. **325:** Constantine convened the **First Council of Nicaea** to try to settle the fever-pitch controversy surrounding the doctrine of **Arianism**. [4][5] The Orthodox held the Son was co-eternal (consubstantial) with the Father. The Arians held that Christ was divine, but that he was not co-eternal with the Father and that he had a beginning. This First Council of Nicaea did not restore Christian unity but did take the first steps that ultimately led to the final form of the **Nicene Creed,** approved at the **Council of Constantinople** in 381, which led to the formalization of the Orthodox position. [6] As for Antioch itself, divisions persisted. After Nicaea, the orthodox authorities in Antioch sent Arius into exile. In 330, the Arian majority of the city retaliated and exiled the city's Orthodox Bishop. This led to serious civil disorder among the Arian and Orthodox Christian rivals in Antioch that involved the whole city. At one point this disorder compelled Constantine to dispatch troops to the city to restore order. [7]

3. **337-361:** Upon Constantine's death in 337, his middle son, Constantius, assumed control of the Eastern Empire and embraced Arianism. By 350, both of his brothers had died leaving him in sole control of the entire Roman Empire. In Antioch in 357 Arians took control of the previous Orthodox Golden Basilica, and the city became a stronghold of Arianism. [8] There is textual evidence during this general time relating to the existence of an image-bearing icon. In his Catechesis, Theodore of Mopsuestia, a native of Antioch, spoke of deacons spreading linens on the altar and representing the figure of the linen cloths at the burial *"so that we may think of him on the altar as if he were placed in the sepulchre after having received the passion."* As mentioned above (see Item H3), the *Sermon of Athanasius*, ascribed to the Bishop of Alexandria (ca. 328-373), recited a church tradition, undoubtedly fashioned in accordance with the precepts of the *Discipline of the Secret*, holding that a full-length body image of Jesus made of boards had been moved from Jerusalem in 68 and was, thereafter, conveyed to Syria. That sermon went on to claim that certain Jewish leaders had driven nails through the image's hands and feet, struck its head with a reed, and pierced its side, causing a large quantity of blood mixed with water to burst forth – wounds intriguingly reflective of those that appear on the Shroud image. In 361 Constantius died and was succeeded by his pagan cousin, Julian, called "the Apostate."

4. **362:** During Emperor Julian's visit to Antioch on October 22, 362, a fire struck the pagan Temple of Apollo, damaging its roof and a statue of the god Apollo. Without proof, the emperor blamed Christians and ordered the Great Cathedral closed and its ecclesiastical treasures confiscated. Before they could be confiscated, however, an Arian presbyter by the name of Theodorus hid the church's treasures. The noted professor Gustavus Eisen records that Theodorus suffered execution rather than reveal an important secret that *"**referred to the treasure which he had hidden and whose hiding place he refused to divulge**."* [9]

In light of subsequent events it would appear that the ecclesiastical treasures included the Shroud. First, during the reconstruction of Antioch in 528-538, an "***awesome image of Christ which was an object of particular veneration***" appeared in the district adjacent to the city's Gate of the Cherubim (see Item H7).

	Secondly, in 945, the *Narratio de Imagine Edessena* related that the *Image of Edessa*, recently brought to Constantinople, had once been hidden in a wall niche located above a city gate where it was found centuries later (see Item H14). Markwardt has suggested that this Byzantine narrative is in error because it identifies the event as occurring in Edessa. He holds that it actually occurred in Antioch where the Shroud in, 362, was concealed in a wall niche above the Gate of the Cherubim. There is another interesting piece of forensic evidence that may also be related to this period of the hidden Shroud: In 2002 Shroud researchers Aldo Guerreschi and Michele Salcito presented an important paper demonstrating that the Shroud has a pattern of water stains consistent with the Shroud being folded and stored in an ancient jar for a prolonged period of time (see Item L10).  Water in bottom of jug is theorized to have wicked up into Shroud leaving water stains **(Fig. 7) Illustration of how Shroud may have been folded and stored in an ancient clay jar**
H7	**363 – 538 AD: Post Constantine Era** 1. **363-410:** When Julian "The Apostate" died in 363, the imperial throne reverted to an orthodox emperor, and in 380 Emperor Theodosius I established orthodox Christianity as the official religion of the empire, condemned the Arian heresy, expelled Arians from Antioch, and restored custody of the Golden Basilica to the Orthodox Melkites. The final form of the **Nicene Creed** soon followed from the **Council of Constantinople** in 381 that formalized the Orthodox position. Nevertheless, the controversy over the nature of the Trinity and the resulting schism of Arianism plagued the Christian Church during the remainder of the 4th century. However, the 5th century produced another divisive controversy, this one over the relationship of the divine and the human natures in Christ. This controversy brought division in the Eastern Church, particularly in Antioch. At one point the controversy led to division of the Christian community in Antioch into four rival sects, with each sect having its own bishop. Ultimately in 451 the **Council of Chalcedon** [1] was convened to resolve the theological issue in favor of the doctrine "*That Christ is one in two distinct natures.*" The council, unfortunately, led to a lasting schism. Many in the East perceived the Council's Christological definition to be heavy handed and to ignore subtle theological issues. Eastern Church representatives blamed the representatives of the Western Latin Church for the division. The Roman Pontiff in Rome, Pope Leo who did not attend the synod, had wanted more time to work out a unified theological definition because  he feared the political fracture lines of the day promoted schism. Unfortunately, his hope was not to be realized. After Chalcedon, Alexandria went into schism (forebearers of today's Copts in Egypt), as did most Christians in Syria, including Antioch. The region was left with three competing branches of Christianity: the Nestorians, a rather small minority; the Monophysites, [2] the majority who rejected the Chalcedon doctrine; and the orthodox Melkites, who adhered to the Chalcedonian formula. The Melkites also generally supported the imperial government of Constantinople; thus, the divisions were not simply theological. The divisions had a political component. The theological arguments themselves were subtle but great enough when coupled with political sentiments to cause radical division. Some twenty years after the council (ca. 471) Monophysites gained control of the church of Antioch. The Patriarch of Antioch fell out of communion with both Rome and Constantinople, and persecution of Antiochene Monophysites continued through the patriarchy of Ephraemius (ca. 528 -545). The only hints of an image archetype during this time are examples of art from the Theodosian era (370-410) that depict Christ with Shroud-like qualities: long and narrow face with long hair parted in the middle and a medium-length beard. [3] **(Fig. 8) From Theodosian era** 2. **525:** In 525 a great fire ravaged Antioch. Soon after, in 526 and 528, major earthquakes struck the city killing more than 250,000 people and destroying almost all of the city's walls and buildings. [4] The Emperor

Justinian financed a great reconstruction project that was carried out over the course of the next decade. Coincidental with Antioch's reconstruction, St. Symeon Stylites the Younger had a vision of Christ appearing on the old city wall that was located near the Gate of the Cherubim. [5] The renowned historian of Antioch Glanville Downey also documents that, at that time and in the same area, there was "**an image of Christ – whether a statue or other representation is not clear from the Greek term eikon that is used to describe it – which was an object of particular veneration.**" [6] The monk John Moschos described the image as "*awesome*." [7] The image's appearance in this location is consistent with the hypothesis that in the year 362 the presbyter Theodorus hid the Shroud and other ecclesiastical treasures of Antioch within a niche of the city wall located above the Gate of the Cherubim (see Item H6), and that these sacred items were rediscovered during Justinian's reconstruction project. [8]

H8 540 AD: Antioch Invaded and Destroyed by the Persians

In June of the year 540 King Chosroes I of Persia (also rendered Khosrau or Khasraw) invaded Syria and marched his army toward Antioch. Chosroes' assault on Antioch resulted in the city being sacked and burned. After the Persian attack, there was never again a reference to the presence in the city of "*an image of Christ …which was an object of particular veneration.*" The destruction of Antioch, the largest city in Syria and the third largest of the great cities of the Byzantine Empire after Alexandria and Constantinople, was essentially complete. It was said that:

> ". . . those few who had not been killed or carried away as slaves could not find the site where once had stood their homes." [1]

One event stands out from those days in the summer of 540. Shroud historian Jack Markwardt and the preeminent historian of Antioch, Glanville Downey, have pointed out that the patriarch of Antioch left the city in the face of the invading Persians and went into Cilicia, an area located on the southern coast of the modern country of Turkey. Both Markwardt and Downey attest that the patriarch **Ephraemius** would not likely have fled in fear; instead, they propose he may have been undertaking an important mission. [2][3]

Some background is important concerning the Patriarch Ephraemius. In the sixth century, there were a number of appointments to high ecclesiastical office of prominent laymen chosen from the ranks of the army and/or the imperial civil service. Ephraemius, the patriarch of Antioch from 527 to 545, was one of these "warrior bishops". An early inscription attests that at some time in his earlier career he was "*comes sacrarum largitionum,*" the head of the central treasury of the entire Byzantine Empire. After this he became *comes Orientis,* the Byzantine administrator of the eastern area of the Empire. Downey reports the following concerning Ephraemius:

> "*. . . he held this office at least from early 523 until shortly before he was named patriarch of Antioch. As 'count of the East' for the empire he was the Byzantine administrator for Palestine and Syria. His office was a peculiarly exacting one, for in addition to the duties which all such posts carried with it, Ephraemius was responsible for the administration of Antioch, where he had his headquarters.*" [4]

Abandoning his post and his flock in Antioch in the face of the enemy would not have been in keeping with the character of this man. Markwardt states that, "*in determining Ephraemius' motive for leaving Antioch, three attendant circumstances must be taken into consideration. First, his principle concern was preservation of Church property.*" [5] In fact, Downey suggests that Ephraemius brokered a deal with Chosroes to spare the great church edifice in Antioch in exchange for the treasures contained therein. Indeed, when Chosroes sacked the city he gave orders to preserve the great church. According to Procopius, a contemporary scholar and historian from the region of Caesarea, Chosroes gave orders to burn everything else. The second consideration suggested by Markwardt is that if Ephraemius had actually abandoned his flock in fear for his own safety, he could not have resumed his patriarchal duties in Antioch, which in fact he did. Third, Markwardt says, "*it is obvious that his departure from the city was deemed entirely appropriate by the emperor and the surviving members of the Antioch church."* [6]

ACHEIROPOIETA: The Greek word *acheiropoieta* (singular *acheiropoieton*) [7] first came into use a short time after the fall of Antioch. The word literally means "**NOT MADE BY HUMAN HANDS**". The designation would first be used to refer to **two** specific images of Christ. Renowned Byzantine art historian Ernst Kitzinger has written about the **"striking development"** of the use of the term *acheiropoieta* to refer to the *Image of Camuliana* (aka, *Image of God Incarnate*) and the *Image of Edessa* "**at almost exactly the same time.**" [8]

| H9 | **540- 692 AD: The Image of God Incarnate – The FIRST to be designated Acheiropoieta** |

1. **540 An Image of Christ in Cilicia:** Shroud historian Jack Markwardt has marshaled evidence to support the hypothesis that the "important mission" undertaken in the year 540 by the Patriarch Ephraemius was to leave the besieged city of Antioch and to go into Cilicia. His mission was to carry to safety a treasured object belonging to the Church of Antioch. [1] Further, it is hypothesized that he conveyed the object to orthodox Cilician churchmen for safekeeping until such time that the Church of Antioch could reclaim it after the city was partially rebuilt, repopulated and made militarily defensible. [2] But, before those conditions could be satisfied in Antioch, Ephraemius died. The year was 545. No retrieval by the church of Antioch of any image that might have been taken into Cilicia ever occurred. Nine years later in 554 a group of orthodox priests publicly paraded an image of Jesus impressed upon linen throughout Cilicia and Cappadocia. [3] This image became the **very first** [4] in all of history to be called **acheiropoieta (not made by human hands)**.

2. **ca. 550 Christ Pantocrator:** The prototype painting of the *Christ Pantocrator* icon (see page 2) first appeared shortly after the destruction of Antioch and, almost simultaneously, with the emergence of the acheiropoieton image in the Byzantine area of Cilicia. The word "Pantocrator" is Greek and means "**Ruler of All**". The oldest known example of the *Christ Pantocrator* icon was discovered in 1962 at Saint Catherine's Monastery in the remote Sinai desert. When the icon was first investigated in 1962, it was covered with a thick yellowish varnish. The icon was carefully restored and the details of its restoration were published in 1967. [5] Subsequent to its restoration the icon was dated to ca. 550 and is considered the oldest of the Pantocrator icon type. The renowned German art historian Hans Belting has stated the following about the *Christ Pantocrator* icon:

> "*. . . it apparently reproduces a well known original of the time that determined the type of Christ preferred in Byzantine painting…the icon's general appearance, in fact, is derived from a concrete model whose identity still is an open question. For all its spontaneity of expression, it was not invented by its painter but seems to reproduce a famous image of Christ that, for this purpose, was replicated for a given commission.*" [6]

What then is the ultimate archetype for the Pantocrator icon? Shroud researchers Mary and Alan Whanger have conducted studies that suggest the key to identifying the archetype for the St. Catherine's Pantocrator is the congruence between the icon and the Shroud of Turin. The Whangers used a process known as the "Polarized Image Overlay Technique" [7] to analyze the congruence between the two images. Their research found over one hundred and fifty (150) points of congruence (PC). Generally, forty-five to sixty PC are enough to declare forensically that two facial images belong to the same person. They concluded:

> "*. . . the Christ Pantocrator icon from Saint Catherine's Monastery is by far the most accurate non-photographic representation of the Shroud image that we have seen.*" [8]

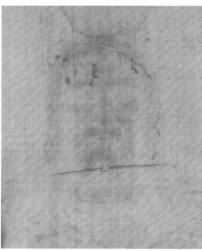

(Fig. 9) The Shroud face (SI)

Along the same lines, the early 20th century French scientist Paul Vignon proposed fifteen markings that could be used to detect possible artistic ties to the Shroud. Later, the Shroud researcher Heinrich Pfeiffer, S.J., culled Vignon's list down to five facial image characteristics that could be used to suggest the Shroud as an archetype for any artistic Christ image. He called his list of characteristics "spy" elements. [9] They are:

(1) Wide space without imprint between cheeks and hair.
(2) Beard slightly displaced to one side.
(3) The moustache is not symmetrical and falls below the mouth at different angles to the left and right of the face.
(4) Possible imprint on the forehead mirroring the blood flow on the shroud.
(5) One cheek is swollen so that the face appears to be slightly asymmetric.

(Fig. 10) Close-up of asymmetrical face of Pantocrator

The artist worked a theological message into the icon based on the asymmetrical face. The message that the artist portrays lies at the heart of orthodox Christology, the area of theology devoted to explaining the nature of Christ. When viewing the icon, the left side of the asymmetrical face shows Christ with a gentle gaze and his hand raised in blessing and mercy that is extended to all of humanity…**Savior**. The right side of the asymmetrical face of the icon shows Christ with a severe expression and a penetrating gaze as he holds the **Book** that contains the Law…**Judge and "Ruler of All."**

(Fig. 11) Savior   **(Fig. 12) Judge and "Ruler of All"**

3. **574 The Image in Cappadocia Seized by the Byzantine Emperor.** In 574 the image-bearing linen cloth that had been displayed in Cilicia and Cappadocia was seized by the Byzantine emperor and taken to the capital city of Constantinople. The "story" given was that the cloth was taken from the remote and tiny village of Camuliana. This cannot be verified. The "story" by the Byzantines that the cloth had come from an insignificant place, the tiny village of Camuliana, may have been made intentionally to blur the image-bearing cloth's true source. If the *acheiropoieton's* true provenance had been acknowledged as being Antioch, the church of Antioch would surely have demanded its immediate return. This was not done; thus, the Byzantines effectively subverted the historical claim that the *acheiropoieton* had direct ties to Antioch and the apostolic era. Later historians were to refer to this *acheiropoieton* as the **Image of Camuliana.** Once the image-bearing cloth safely arrived in Constantinople, the Byzantines always used the name **Image of God Incarnate**. [10][11][12]

4. **586 Image of God Incarnate used as a Palladium (protective image):** The *Byzantine Chronicle* (ca. 625) written by the historian Theophylact Simokattes reports that the Byzantine general, Philippikos, used the facial portion of the *Image of God Incarnate* as the model for an army palladium in the year 586. Simokattes reported that the image that had been placed on an *Imperial Labarum* (Byzantine military standard) was *"... stripped of its sacred coverings and paraded through the ranks, thereby inspiring the army with a greater and irresistible courage."* [13] An additional historical mention of the *Image of God Incarnate* being deployed as a palladium describes how in the year 626 the image was deployed against the Avars, who were besieging the capital city of Constantinople. The use of the palladium helped rally the forces of the capital, and the barbarian Avars were repelled. Although Byzantine Imperial and Church authorities would know about a full-body image of a beaten, crucified and naked Jesus, they had several compelling reasons to promote the image as that of a living and triumphant Jesus. First, Christian art objected to stripping Christ of his garments. [14] Crucifixes and crucifixion portrayals were invented just after the arrival of the *Image of God Incarnate* in Constantinople. The example shown below is from the Syrian Gospel Book, dated 586, that shows Jesus wearing a robe or *colobium*: [15] This is the first known depiction in an illuminated manuscript of the crucifixion of Christ.

(Fig. 13) Rabula Gospels, ca. 586

Thus, Byzantine concepts of modesty at this time would have precluded making artistic renditions of the *Image of God Incarnate* that depicted a naked Christ. Second, Emperor Justin II who seized the image-bearing cloth from Cilicia in order to use the image as a palladium, did not wish to portray Jesus as scourged, naked, crucified and dead. He desired to use the image to make the Byzantine capital and empire **Theophylaktos**; that is, protected by God himself. Thus, imperial authorities had to present the image as one of a **triumphant** Christ, devoid of any signs of injury, in order to engender public confidence in their ability to provide them with perpetual divine protection. Third, the Byzantines were concerned, because of their piety, how God might react to imperial exploitation of a holy object. They placed copies only of the image's face, neck, shoulders, arms, hands and upper torso, absent all injuries, upon their military standards, or **labara**.

5. **589 Mozarabic Rite:** About the year 589 the Visigoth Church of Spain began to recite in their version of the liturgy (the Mozarabic or Rite of Toledo) [16][17][18] the following statement as part of the Offertory for the first Saturday after Easter:

> *"Peter ran with John to the tomb and saw the recent imprints (vestigia) of the dead and risen man on the linens."* [19]

This is yet another written reference known to imply the existence of a full-length body image of Christ. As noted, other such references are found in the *Epistle to the Galatians*, the *Inscription of Abercius*, the *Hymn of the Pearl*, and the *Sermon of Athanasius*. Saint Leander, [20] the Bishop of Seville, is widely credited with the Mozarabic text. History has also largely credited him with the conversion of the Arian Visigoth kings of Spain to orthodox Nicene Christianity. In 589 Leander convoked the Third Council of Toledo and delivered the triumphant closing sermon that marked the conversion of the Visigoths. [21] What is most significant is that, from 579 to 582, he was in Constantinople, having been exiled by the then Arian Visigoth king, Liuvigild. While in Constantinople Leander became good friends with Gregorius Anicius, who was the representative of Pope Pelagius II to the Byzantine court. Gregorius, the friend of Leander, would later become one of the most famous of all popes, **Gregory the Great** (pontificate: 590-604). [22] The future Gregory the Great had privileged access as an insider in the Imperial Court. This access would have given him knowledge of the *Image of God Incarnate,* and it is conceivable he may have been one of the very few, as the representative of the pope to the Imperial Court, to gain access to actually view the acheiropoieton. When he was pope, he imported to Rome a tempura painting of Christ on wood that he installed in the Sancta Sanctorum of the Lateran Palace. He named this work the **Acheropita**, thereby clearly denoting that it reflected the *acheiropoieton* image in Constantinople – the *Image of God Incarnate.* Leander himself would not have had the access to the image that Gregorius did. He could only have come to "know" the truth of the "*recent imprints*" from his close friend, the future Gregory the Great.

The Mozarabic Rite is not the first reference to Christ's burial Shroud in the rites of the Church. From the earliest days of Christianity to the present day, the Catholic Church has provided a tangible reminder of the linen burial Shroud of Christ at every one of its Masses through the use of the corporal linen. The word "*corporal* " comes from the Latin word *corpus* (*corporis*) meaning body. Thomas Aquinas in his *Summa Theologica* (written 1265-1274) states:

> "*. . . yet the corporal is made of linen, since Christ's body was wrapped therein. Hence we read in an Epistle of Pope Sylvester (pontificate: 314-335, during the era of Constantine the Great), quoted in the same distinction: "By a unanimous decree we command that no one shall presume to celebrate the sacrifice of the altar upon a cloth of silk, or dyed material, but upon linen consecrated by the bishop; as Christ's body was buried in a clean linen winding-sheet.*" [23] (cf. St. Thomas, *Summa Theologica* III, q.83, a.3, ad.7)

6. **692 Justinian II Solidus Coin:** In the year 692, the Byzantine Emperor Justinian II (reign 685-695 and 705-711) convened a church council in Constantinople. The council was convened without papal authority from Rome, so it is not considered as one of the ecumenical councils. The council was held in the Trullan hall of his great palace, and hence, the council became known as the **Council of Trullo** (aka **Quinisext Council**). [24] Canon 82 of the council does appear to be have been singled out for acceptance years later in a letter by Pope Adrian I (pontificate 772-795) to the Patriarch of Constantinople named Tarasius, who is today recognized as a saint by both the Orthodox and Catholic Churches. Canon 82 states that Jesus is no longer to be represented simply as a lamb but in human form so "...*that we may recall to our memory his conversations in the flesh, his passion and salutary death, and his redemption which was wrought for the whole world.*" [25]

Almost simultaneously with the publishing of the canons of the Trullo council, the Emperor Justinian II minted the first official Byzantine *solidus* coin with a facial image of Christ. Constantine I had first introduced the *solidus* coin in 309-310, and this type of gold coin was used throughout the Eastern Roman Empire (Byzantium) until the tenth century. The solidus replaced the *aureus* as the main gold coin of the empire. The word "*soldier*" is ultimately derived from *solidus,* as this is the coin type used for the pay of the Roman and Byzantine military. The earliest of the Justinian II *solidus* coins depict Christ in frontal position with a cross behind His head. He has long, wavy hair, a beard, and a mustache. He is bestowing a blessing with His right hand and holds the Book of the Gospels in His left hand, similar to the Pantocrator icon. Written around His head are the words "**Christ, King of those who rule.**" On the reverse side of the coin there is an image of the emperor with the words "**Lord Justinian, the servant of Christ**".

In 2015 Giulio Fanti and Pierandrea Malfi co-authored an important book entitled "*The Shroud of Turin: First Century after Christ!.*" The book includes a long and detailed chapter devoted to the numismatic investigation of the Justinian II *solidus coin* minted in 692, as well as other coins bearing an image of Christ. The authors provide an in-depth presentation of the tight correlation between the Shroud and the numismatic characteristics of the *solidus* coin. Their study includes an exacting evaluation of an extensive list of "coincidences" that echo and build on the Vignon and Pfeiffer characteristics. They performed a statistical evaluation on the whole set of "coincidences" and report in their study that their statistical calculations returned a certainty **greater than 99.99%** that the Shroud was the model for Justinian's 692 gold solidus coin. [26]

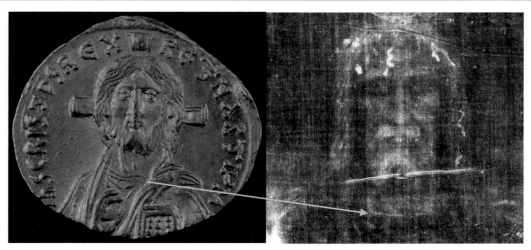

(Fig. 14)   Justinian II Coin 692 AD

Shown above is the 692 Justinian II solidus coin with a photographic negative of the Shroud face **(SI)**. The negative image, which was not available until the year 1898, could not have been the archetype for the coin. The actual faint Shroud image must have been the archetype for the coin. However, the negative shows an interesting "macro-characteristic" of the Shroud that is visible only on close visual inspection of the actual Shroud cloth. In the negative it is an easily observed "characteristic" illustrating the detail and care that must have been taken by the coin engraver, along with the intimate access he must have been given by the Emperor Justinian II to the actual cloth. The arrow points out this feature: It is a subtle double fold in the cloth just below the neck. In their book, Fanti and Malfi have hypothesized that this double fold is interpreted on the coin as the hem of Jesus' garment.

H10	**The Image of Edessa – The SECOND to be designated Acheiropoieta**

The city of Edessa, today known as the Turkish city of Urfa, is approximately 145 miles northeast of Antioch. At the beginning of the Christian era Edessa lay in the Parthian, not the Roman sphere of control, and its people spoke Syriac not Greek. [1] Nevertheless, the city was a natural target for early evangelization, primarily because the city had a significant Jewish population. Christian missionaries relied on the friendship of the Jews to successfully evangelize. The story spread that Christianity became the dominant faith in the city and that Edessa was the first kingdom to adopt Christianity as its official religion. [2] In particular, the story of the conversion of King Abgar V was written down and widely circulated. [3]

1. **ca. 325 Eusebius writes his Church History:** Eusebius, known as the "Father of Church History", in his famous ***Historia Ecclesiastica*** (Church History), writes about the conversion of King Abgar V who ruled the city of Edessa in the first century from AD 13-50. [4] The legendary story of Eusebius reports that Abgar V was seriously ill and sent a written message inviting Jesus to travel to Edessa to cure him and to teach his people. It is reported that Jesus sent a return letter promising to send a disciple to the city. The tradition, according to Eusebius, is that Jesus' disciple Thomas (Didymus) sent Thaddeus (Addai) to Edessa. Many historians judge the story to be apocryphal, but the legend of the correspondence between Abgar and Jesus became famous throughout Christendom.

2. **ca. 190 The conversion of Abgar VIII (the Great) of Edessa to Christianity:** The conversion of Abgar VIII (the Great), along with the possibility of Abercius taking a full-body image of Jesus from Antioch to Edessa in support of that conversion, and the associated *Hymn of the Pearl*, are documented above (see Item H5). Even a vague "memory" of a full-body Christ image in Edessa, associated with the later conversion of Abgar VIII (the Great), would likely come to be appropriated and incorporated into the original Abgar V conversion legend initially authored by Eusebius.

3. **ca. 400 Doctrine of Addai:** This is a second Syriac Christian text, after Eusebius, that speaks about the conversion of Abgar V in the first century. [5] In this text, reference is first made to an image, a painting of Jesus made with "***choice paints***", being instrumental in the conversion of Abgar V. There is no mention concerning the conversion of Abgar VIII (the Great) in 190, but mysteriously, the narrative of a portrait enters the story, along with the story of correspondence between Abgar and Jesus, that was first reported

by Eusebius. It is stated quite clearly that the king's archivist and artist, Hannan painted the image. [6] In the *Doctrine of Addai*, the "image" played a relatively minor role. Nevertheless, over time allusion to the portrait gradually increased in sanctity and importance. The foremost historian of Edessa, Judah B. Segal, has stated that, *"In the earliest version, it was the work of the painter Hannan, in later accounts it would be painted only with the assistance of Jesus, finally it was wholly the work of Jesus himself."* [7]

4. **544 Siege of Edessa:** Just four years after the sack and destruction of Antioch in 540, Chosroes and his Persian army turned north to besiege the city of Edessa. In a close fight the citizens of Edessa repulsed Chosroes' army. Evagrius Scholasticus, writing in his *Ecclesiastical History* (ca. 590), described the battle for the city of Edessa and relates how an image of Christ of "*divine origin*" was given credit by the people of Edessa for their victory. [8]

5. **ca 590-593: "Acheiropoieta" designation for the Image of Edessa:** The designation ***acheiropoieta* (not made by human hands)** was given to the *Image of Edessa* around the year 590-593 by the Church historian, Evagrius Scholasticus. [9] The *Image of Edessa* thereby became the **second** image to be given this designation, joining the only other image of Christ with this designation, the *Image of Camuliana* (aka, the *Image of God Incarnate*), which was given that designation by the Church historian, Pseudo-Zachariah, in 568-569. [10] The *Image of Edessa* would now carry the reputation of being "wholly the work of Jesus himself." [11] The image was soon to become famous throughout Christendom with its legends known and venerated throughout Western Europe and Byzantium.

6. **639 Muslim Conquest:** In 639 Edessa fell under Muslim control. At the time of the conquest the three main sects of Mesopotamian Christianity, the Orthodox Melkites, the Nestorians and the Monophysites were all represented in the city. [12] All three claimed to have possession of the true *Image of Edessa*, and there was great rivalry among the sects. The hatred of the Melkites by the majority Monophysite community in Edessa outweighed even their fear of the Muslims. In Edessa, the cloth apparently was always kept in a frame that never revealed it to be anything more than a cloth bearing a facial image of Jesus. By the time of the Muslim conquest, the *Image of Edessa* was, uniquely, the only significant "icon" of Christ that had not been appropriated in the name of the Byzantine emperor and taken to Constantinople. [13]

(Fig. 15)   Abgar V receiving image (10th Century)

H11	**ca. 614-711 AD: The Sudarium of Oviedo**
	The Gospel of John mentions a second cloth seen in the tomb on Sunday morning after the crucifixion, death and burial of Jesus of Nazareth. [1]

So Peter and the other disciple went out and came to the tomb. They both ran, but the other disciple ran faster than Peter and arrived at the tomb first; he bent down and saw the burial cloths there, but did not go in. When Simon Peter arrived after him, he went into the tomb and saw the burial cloths there, and the cloth that had covered his head, not with the burial cloths but rolled up in a separate place.

John 20: 3-7

What then is the explanation for the *"cloth that had been on Jesus' head?"* The most common interpretation is that this was a cloth that had covered Jesus' face after he died on the cross and preserved his dignity as he was lowered from the cross and transported from the site of crucifixion to the tomb. Covering the face with a cloth would have been in accord with Jewish sensitivities. The cloth would likely have become soaked with the blood and bodily fluids of Jesus and would therefore, according to Jewish burial requirements, have been placed in the tomb with the body.

In approximately the year 614 a cloth was carried out of the east, possibly from Syria or Palestine, through Alexandria, Egypt and then across North Africa or by sea west across the Mediterranean toward Spain. The cloth apparently was continually moved ahead of conquering Persian forces, ultimately arriving in the city of Oviedo, Spain, where it remains to the current day. The exact date of the arrival of the cloth in Spain is unknown, but certainly before the invasion of the Iberian Peninsula by Islamic forces in the year 711.[2] The cloth that arrived in Spain is known today as the **Sudarium of Oviedo**. It is a blood-stained cloth measuring 33" x 21" (84 x 53 cm) in size.

Forensic evidence supports the conclusion that the Sudarium shares bloodstains that can be mapped to the Shroud of Turin [3][4][5]. This lends support to the claim that the Sudarium is the same face cloth that covered Jesus' face as he was carried from the place of his crucifixion to his tomb – the cloth that John's Gospel says Peter saw "***rolled up in a place by itself***". It is historically certain that the Sudarium has not been in contact with the Shroud since its arrival in Spain ca. 614-711. The Shroud of Turin has a known history that lies only **north** of the Mediterranean. The only place the Sudarium and Shroud, if both are authentic, could have picked up matching bloodstains from the same body is in Palestine prior to the year 614.

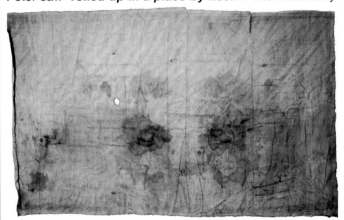

(Fig. 16) Sudarium of Oviedo

In 2012 X-ray fluorescence testing, a technique widely used for elemental and chemical analysis, was performed on the Sudarium by a team of researchers. The testing was authorized by the custodians of the Sudarium and was done at the Oviedo Cathedral where the Sudarium is preserved. On March 9 and 10, 2012, fifty-seven tests based on a 2x2 cm grid layout of the entire Sudarium were performed. It was found that the highest content of dirt containing calcium was observed close to the area corresponding to the tip of the nose, based on mapping of the congruence of bloodstains found on the Shroud and Sudarium. The research team subsequently obtained dirt samples from the Calvary site in Jerusalem. They found that the chemical signatures of the dirt on the Sudarium could be closely correlated with that of the Calvary samples, circumstantial evidence that it was once in that same region [6] (also see ItemL11.0).

H12	**711-943 AD: Iconoclasm and the "Covenant with God"**

1. **711:** The Emperor Justinian II was deposed and beheaded. Some Byzantine Christians may have interpreted his fate as divine punishment for his having placed images of Jesus upon imperial coinage. The iconoclast doctrine that originated in Judaism held that religious images might constitute idolatry or encourage profane forms of worship. Indeed, by 717 when **Leo III the Isaurian** seized the throne from Theodosios III, the majority of the Byzantine clergy were opposed to the display of sacred images.

2. **717-718:** Constantinople was besieged by a combined land and sea offensive by a large Muslim Arab army less than a hundred years after the death of the founder of Islam, Muhammad the Holy Prophet. Leo III prevailed but it was a fierce battle that the Byzantines won only with "providential" help. The Arab army was ravaged by starvation and infectious epidemics. The Lombard historian, Paul the Deacon, put the number of their dead from starvation and disease at 300,000. [1] Leo III was a moderate iconoclast, but his "providential" victory would only reinforce his commitment to that doctrine. Throughout the battle for the empire Leo III did not resort to using the *Image of God Incarnate* as a palladium as it had been used to defeat the Avars in 626.

3. **726:** A volcanic eruption in the Mediterranean spewed smoke and ash all over Asia Minor. Leo III and clerical authorities interpreted these events as a further sign of divine wrath brought on by idolatrous practices. The emperor moved swiftly to preclude the public display of all religious images.

4. **740:** The Byzantine historian, Theophanes, reports that on October 26, 740, the capital city of Constantinople was struck by a major earthquake which was followed by a long series of aftershocks, some very violent, that "continued for twelve months." [2] For the better part of this twelve-month period, the reigning emperor was Leo III. Leo died on June 18, 741, and his son, **Constantine V**, became emperor (741-75). During Constantine's reign, terrible plagues also afflicted Constantinople and other areas of the Empire. [3] Constantine's interpretation of a natural catastrophe, earthquake, volcanic eruption or plague was reflected in his understanding of such an event as divine punishment or a warning. His position on **Iconoclasm** was much stronger than his father's and was made starkly clear, as he decreed:

> *"He cannot be depicted. For what is depicted in one person, and he who circumscribes that person has plainly circumscribed the divine nature which is incapable of being circumscribed."* [4][5]

This language was used to make a theological point, but intriguingly, at the same time it invokes the nature of the Shroud image itself. There are absolutely no distinguishing borders that can be associated with the Shroud image. As will be seen in the empirical sections that are presented further on in this document, the Shroud image at its periphery simply disappears into the cloth with no fixed border, as would be the case for a typically painted image, let alone some other man-made object such as an engraving or sculpture. Constantine V would go further; he convoked a council in Constantinople, the *Council of Hieria*. This council was never recognized as an ecumenical council, but it did hold sway in Byzantium. The final act of the council took place on 27 August 754, when Constantine and his son Leo, along with the bishops attending the council, processed to the *Forum of Constantine* and read out the acts of the council. Included was the insistence that Christ could not be represented by an image, since this would be to separate the human from the divine. Canon 264C of the council stated: *"**The only true image of Christ is the bread and wine of the Eucharist as he Himself indicated.**"* [6] With this proclamation iconoclasm was institutionalized in the empire.

In 1995, a Byzantine scholar, Krijinie N. Ciggaar, published a French translation of the anonymous Tarragonensis 55 (generally known as the Tarragon manuscript). The document is written in Latin and is maintained in the Public Library of Tarragon, Spain. [7] The document has been confidently dated to 1075-1098, with the most likely date 1081-1098. The document looks back to the time of the decree of Constantine V and that of the Council of Hieria, and includes the following words about the golden case containing the *Image of God Incarnate*:

> *". . . [it] is not shown to anyone and is not opened up for anyone <u>except</u> the emperor of Constantinople. The case that stored the holy object used to be kept open once, but...a heavenly vision revealed that the city would not be freed of such ill until such time as the linen cloth with the Lord's face on it should be locked up and hidden away far from human eyes. And so it was done."* [8]

Shroud historian Jack Markwardt maintains that the Tarragon manuscript is evidence that the Emperor Constantine V made a "**Covenant with God**" that the *Image of God Incarnate* would henceforth be sealed in its golden case away from public view in perpetuity, and that *"**so it was done.**"* Constantine V bound himself and his imperial successors to reserve the image for viewing **only** by the Emperor himself. Constantine V would thereby consign the Shroud to essentially four and a half centuries of historical obscurity. [9]

5. **800**: With the Shroud locked away from public sight in a golden case kept in the Imperial Palace, the references to the extant publicly known *acheiropoieta image* in Edessa became increasingly blurred with facts actually related to the hidden-away *Image of God Incarnate*. It didn't take long after Constantine V's *Imperial Covenant with God* for this blurring to emerge. As early as ca. 800, the so-called "**Latin Abgar Legend**" [10] was published in western circles. In this revised version of the Legend, Jesus tells Abgar V, "*If you wish to see my face in the flesh, behold I send to you a linen, on which you will discover not only the features of my face, but a divinely copied configuration of my entire body.*" [11] The narrator of the story then goes on to record that…"*in order that in all things and in every way he might satisfy this king, spread out his entire body on a linen cloth that was white as snow. On this cloth, marvelous as it is to see or even hear such a thing, the glorious features of that Lordly face, and the majestic form of his whole body were so divinely transferred, that for those who did not see the Lord when he had come in the flesh, this transfiguration on the linen makes it quite possible for them to see.*" [12]

6. **836**: In 836, during the second period of imperially imposed Iconoclasm, the three orthodox Melkite Patriarchs of Alexandria, Antioch, and Jerusalem, are reported to have joined together to draft a remarkable letter addressed to the iconoclastic Byzantine Emperor Theophilus in Constantinople. [13] In their letter they are reported to have set out a list of icons that were "*made without human hands*" (*acheiropoieta*) and petitioned for moderation of iconoclasm. At the head of their list was the *Image of Edessa*. The second iconoclastic period ended shortly after the death of Theophilus on 20 January 842. Just weeks later, on 19 February 842, the first Sunday of Lent, icons were brought back to the churches of Byzantium. This first Sunday of Lent in the Orthodox Church is still celebrated as the "**Feast of Orthodoxy**" (also known as the **Sunday of Orthodoxy** or the **Triumph of Orthodoxy**), a commemoration of the date that the liturgical use of icons was restored. [14] It was also at about this same time that the term **Mandylion** appears to have first been used to refer uniquely to the *Image of Edessa*. There are various theories about the origins of the word. Some scholars think it is derived from the Arabic *mandil* meaning a small cloth-like towel. Still others think it derives from the Latin *mantilium*, a general word for a larger cloth. [15]

H13	**943 AD: Byzantium Captures Image of Edessa**
	In the summer of 943 Byzantine Emperor Romanus I ordered an army of 80,000 to besiege the Muslim-held city of Edessa. The siege was ordered with the single intent of capturing the *Image of Edessa* and securing it for the Byzantine Empire. The Muslim ruler, in an effort to save his city, demanded that the three different Christian sects that were then represented in the city of Edessa give up their respective images to the Byzantines. After protest and rioting all three sects are reported to have surrendered their respective "true" copies of the *Image of Edessa* to the Byzantine forces. It is further reported that the Byzantines carefully studied the various images and then retained only the image that had been in possession of the Orthodox Melkites, while the other two "true" copies were returned to their respective sects. [1][2] With the image securely in their hands the Byzantines lifted the siege of the city with the *Image of Edessa* in their possession.
H14	**944 – 1203 AD: Both Acheiropoieta Images in Constantinople**
	1. **August 15, 944:** On this date the captured *Image of Edessa*, the *Mandylion*, arrived in the Byzantine capital city of Constantinople. As of this date the ***Image of God Incarnate*** and the ***Image of Edessa***, the two *acheiropoieta* images that appeared at "***almost the same time***" [1] in the 550s, shortly after the fall of Antioch, are both in the possession of the Byzantine Emperor in Constantinople. The **Chronicle** [2] of Symeon Magister Metaphrastes reports that in the evening of August 15 the *Image of Edessa* was viewed by the future Emperor Constantine VII and the two sons of the current Emperor Romanus. While the Emperor's sons could only see the face of the image, Constantine could also see the eyes and ears of the "faint" image. This event may have actually transpired, or Symeon may have invented the episode to praise Emperor Constantine's "*innate spiritual qualities.*"

2. **August 16, 944:** This was the day the public welcomed the *Mandylion* to the capital. The archdeacon of Hagia Sophia Cathedral, **Gregory**, held the title of "**Referendarius**", a title given to an officer in the Byzantine Imperial Court who reported directly to the Emperor. Gregory gave a public sermon on the occasion. A surviving text of the sermon includes two descriptive passages. In the first passage, Gregory figuratively quotes Jesus speaking about the image, and in the second he gives his own reflection on the newly arrived *Image of Edessa*: |

Jesus: "*I have put it on my face and have shown that this is the radiance of the face you were seeking.*" [3]

Gregory: "*This reflection, however – may everyone be inspired with the explanation – has been imprinted only by the sweat from the face of the Ruler of Life, falling like drops of blood, and by the finger of God.*" [4]

One year later the **Narratio De Imagine Edessena,** [5] reputedly commissioned by the emperor Constantine Porphyrognitus himself, produced a new "history" of the legend of Abgar V. The **Narratio** confirms the century-old tradition that the *Image of Edessa, Mandylion,* was a cloth bearing a facial image of Jesus:

"*The gospel tells us that his sweat fell like drops of blood and then it is said that he took this piece of cloth, which can still be seen, from one of his disciples, and wiped off the streams of sweat on it. The figure of his divine face, which is still visible, was immediately transferred onto it.*" [6][7]

Markwardt suggests that further blurring occurred in this new version of the "legend" of Abgar V just as it did in the ninth century *Latin Abgar Legend*. Specifically, he suggests the Narratio's version of the *Image of Edessa* being hidden in a wall niche located above a city gate in Edessa is actually a substitution of Antiochene history reported by the historian Glanville Downey of **"an image of Christ . . . which was an object of particular veneration"** that was associated with the city wall of Antioch and the Gate of the Cherubim. [8]

3. **958**: Emperor Constantine VII sent a letter to rally his troops who were engaged in the area of Tarsus, and in his letter he specifically mentions the empire's possession of passion relics including:

"**. . . the sacred linens (σπάργανα), the sindon which God wore, and other symbols of the immaculate passion.**" [9][10]

The Greek word "sindon" refers to a fine, thin fabric of linen. Here Emperor Constantine VII Porphyrogennetos seems to be stating unequivocally that the burial sindon of Jesus, the *Image of God Incarnate*, was still in existence and in the possession of the Byzantine Emperor.

4. **ca.1164**: By the year 1164 a new iconographic image had emerged in Byzantium, the **Epitaphios** or **Threnos.** [11] An early example of the genre from Nerezi, Serbia is shown to the left.[12] Jesus is shown lying on his shroud with his head cradled by his mother. Epitaphios icons from this era point to the knowledge of the existence of a linen cloth archetype in Byzantium that carries a full-length image of Jesus' crucified and dead body. [13]

The tradition of the **Epitaphios** is carried on even today in the the Eastern Orthodox Churches and those Eastern Catholic Churches which follow the Byzantine Rite, as shown in the image below. The image shows how the Epitaphios is displayed in Good Friday and Holy Saturday services.

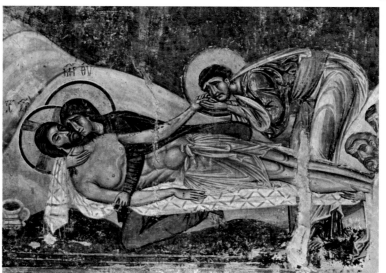

(Fig. 17)   Epitaphios from Nerezi, Macedonia, ca. 1164

| H23 | **1968: New Crucifixion Evidence** [1][2][3] |

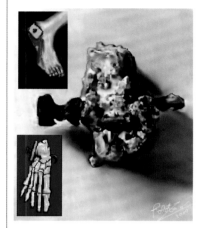

(Fig. 37) Yehochanan's nailed heal bone details

In 1968 a skeletal heel bone, with a 7-inch-long (17.9 cm) spike driven through it, was discovered in an ossuary, or bone box, inside a first-century tomb in the vicinity of Jerusalem. *"The heel, which belonged to a man named Yehochanan, helped settle a long-simmering historical debate about the plausibility of Gospel accounts of Jesus' tomb burial. Crucifixion was a punishment reserved for the dregs of society, and some experts have scoffed at the idea that Romans would accord anyone so dispatched the dignity of a proper internment."* [4] Some *"historical Jesus"* skeptics have even suggested that most likely *"Jesus' remains, like those of other common criminals, would have been left to rot on the cross or tossed into a ditch, a fate that would certainly complicate any resurrection narrative."* [5] The heel bone discovered in 1968 is, of course, not the only "stunner" to be disgorged from the ground in both Galilee and Jerusalem by contemporary archaeology that support the Gospel narrative. [6] But Yehochanan's heel bone is singularly significant from an historical perspective because it directly offers *"an example of a crucified man from Jesus' day for whom the Romans permitted a Jewish burial."* [7]

Now there was a virtuous and righteous man named Joseph who, though he was a member of the council, had not consented to their plan of action. He came from the Jewish town of Arimathea and was awaiting the kingdom of God. He went to Pilate and asked for the body of Jesus. After he had taken the body down, he wrapped it in a linen cloth and laid him in a rock-hewn tomb in which no one had yet been buried. [8]

Luke 23: 50-53

| H24 | **1978: Shroud of Turin Research Project (STURP)** [1][2] |

In 1976 John Jackson, an active duty United States Air Force Officer, and Bill Mottern, a scientist from the Sandia National Laboratory, worked to generate the first 3-dimensional map of the Shroud image. Later in that same year Jackson, while teaching physics at the Air Force Academy, in partnership with Eric Jumper, another active duty Air Force Officer teaching science at the Air Force Academy, used a VP-8 analog image analyzing computer furnished by Pete Schumacher, an engineer with Interpretation Systems, Inc., to make a brightness map of the Shroud image. The resulting brightness map confirmed that 3-dimensional information was encoded in the Shroud image (see Item B3). Jackson and Jumper assembled a team of cadets to assist with model building based on the Shroud brightness map. Today a display exists in the lower level of the Air Force Academy Chapel (Colorado Springs, Colorado) that commemorates this pioneering work that directly led to the organization of the **Shroud of Turin Research Project (STURP)**, a project born at the USAF Academy.

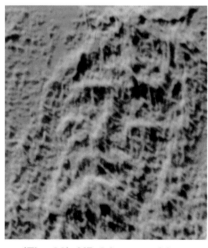
(Fig. 38) VP-8 Image of face

(Fig. 39) U.S. Air Force Academy Chapel

The STURP research team was composed of a large group of outstanding American scientists and support personnel. In October of 1978 the STURP team traveled to Turin, Italy, to conduct an in-depth scientific examination of the Shroud. This American expedition conducted what is still the most extensive hands-on study of the Shroud ever undertaken. The STURP team spent over two years prior to embarking for Turin in planning dozens of specific data-gathering experiments, measurements and tests. To support their efforts, they carried to Turin several tons of equipment, including sophisticated scientific measuring and data-gathering instruments. The team arrived in Turin in early October 1978 following a public display of the Shroud commemorating the 400th anniversary of the Shroud's arrival in that city. For five full days, starting on October 8th, the STURP team examined the Shroud around the clock in a large room at the Royal Palace adjoining the Turin Cathedral. Each 24-hour period was broken down into shifts that allowed the work to proceed uninterrupted while some STURP staff slept and others conducted research.

(Fig. 40) Members of STURP scientific team in Turin

Among the methods used to gather data were direct microscopy, infrared spectrometry, X-ray fluorescence spectrometry, X-ray radiography, thermography, and ultraviolet fluorescence spectrometry. In addition, a broad spectrum of photographic data was collected. Ultraviolet fluorescence photographs, raking-light photographs, normal front-lit photographs and backlit photographs of the entire Shroud were taken, as well as dozens of micro-photographs of strategically selected areas of the Shroud. The STURP team also collected sticky tape samples from the surface of the Shroud cloth as well as thread samples that were retained and returned to the United States for further studies. Subsequent studies of these samples were conducted using microscopy, pyrolysis-mass-spectrometry, laser-microbe Raman analysis and various methods of micro-chemical testing. The results of STURP research were published in twenty (20) peer-reviewed scientific journal articles over the four years following the team's conclusion of work in Turin. [3] In addition, numerous other papers have subsequently been published, elaborating on findings and data from the STURP expedition.

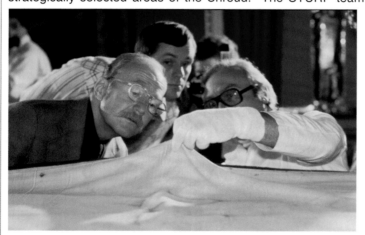

(Fig. 41) Chemist Ray Rogers (left), and physicist John Jackson (behind) examine an area of the backside of the Shroud during the STURP expedition, an area not seen for more than 400 years.

In October 1981, the final official report of the STURP team effort was issued. At a press conference to mark the occasion the following official summary of the STURP conclusions was handed to the press: [4]

No pigments, paints, dyes or stains have been found on the fibrils. X-ray, fluorescence and microchemistry on the fibrils preclude the possibility of paint being used as a method for creating the image. Ultraviolet and infrared evaluation confirm these studies. Computer image enhancement and analysis by a device known as a VP-8 image analyzer show that the image has unique, three-dimensional information encoded in it. Microchemical evaluation has indicated no evidence of any spices, oils, or any biochemicals known to be produced by the body in life or in death. It is clear that there has been a direct contact of the Shroud with a body, which explains certain features such as scourge marks, as well as the blood. However, while this type of contact might explain some of the features of the torso, it is totally incapable of explaining the image of the face with the high resolution that has been amply demonstrated by photography. The basic problem from a scientific point of view is that some explanations, which might be tenable from a chemical point of view, are precluded by physics. Contrariwise, certain physical explanations which may be attractive are completely precluded by the chemistry. For an adequate explanation for the image of the Shroud, one must have an explanation which is scientifically sound, from a physical, chemical, biological and medical viewpoint. At the present, this type of solution does not appear to be obtainable by the best efforts of the members of the Shroud Team. Furthermore, experiments in physics and chemistry with old linen have failed to reproduce adequately the phenomenon presented by the Shroud of Turin. The scientific consensus is that the image was produced by something which resulted in oxidation, dehydration and conjugation of the polysaccharide structure of the microfibrils of the linen itself. Such changes can be duplicated in the laboratory by certain chemical and physical processes. A similar type of change in linen can be obtained by sulfuric acid or heat. However, there are no chemical or physical methods known which can account for the totality of the image, nor can any combination of physical, chemical, biological or medical circumstances explain the image adequately.

Thus, the answer to the question of how the image was produced or what produced the image remains, now, as it has in the past, a mystery.

We can conclude for now that the Shroud image is that of a real human form of a scourged, crucified man. It is not the product of an artist. The bloodstains are composed of hemoglobin and also give a positive test for serum albumin. The image is an ongoing mystery and until further chemical studies are made, perhaps by this group of scientists, or perhaps by some scientists in the future, the problem remains unsolved.

H25 — 1983: New Custody for the Shroud

On March 18, 1983, Umberto II of the House of Savoy, the deposed last King of Italy, died in exile at his home in Portugal. Before he died he had made arrangements that upon his death the official ownership of the Shroud would be passed to the **Holy See**, the seat of the pope as the Bishop of Rome. [1] At the time of Umberto's death the Shroud had been in the possession of the House of Savoy for 530 years and in the city of Turin for 405 years, except during World War II, when Umberto II's father King Victor Emmanuel III ordered that the Shroud be moved out of the city for its safety. Accordingly, the Shroud was moved to the Benedictine Abbey of Montevergine which lies just to the north-east of Naples in the province of Avellino. [2] For seven years the Shroud was housed secretly at the Abbey with only the prior of the monastery, along with the vicar-general and two of the monks, entrusted with the knowledge of the Shroud's presence. After the war in 1946, the Shroud was again moved back to Turin and has remained there since, but now under the ownership of the Holy See and under the care of the designated **Papal Guardian** of the Shroud, the Bishop of Turin.

(Fig. 42) The Abbey of Montevergine, the World War II protective home of the Shroud

(Fig. 43) John and Rebecca Jackson discuss Shroud research with Pope John Paul II (1997)

H26	**1988: Radiocarbon Dating** The custodian of the Shroud, the Archbishop of Turin, authorized a sample to be cut from one corner of the Shroud for radiocarbon dating. At a press conference on 13 October 1988, the results were announced: The Shroud linen cloth was declared to date from 1260 to 1390 AD. [1][2][3][4] Dark clouds gathered around the Shroud. The door slammed shut and was bolted in the eyes of the greater scientific community. Skeptics appeared to be vindicated. The public turned away. Like the kenosis, the self-renunciation of the divine nature of Jesus Christ himself in the incarnation, the kenosis of the Shroud began…. a tragic period of emptying from public view and interest (see extended discussion on the *Dating of the Shroud* in Section 7).
H27	**2002: Shroud Preservation Project** In early 2000, the Turin custodians of the Shroud hosted a symposium to consider, among other issues, the conservation of the Shroud. [1] A proposal was subsequently drafted that recommended intervention for two primary reasons: 1. With the passage of time, the stitching that secured the Holland backing cloth (see Item H19) and the patches that covered the burn damage from the 1532 fire were judged to be causing stress that deepened various creases on the Shroud. 2. The burned and blackened char material beneath the patches that covered the burn holes from the same 1532 fire was also thought to be acidic in nature, and it was feared the char might be slowly eroding the back of the Shroud. It was also feared that loose char particles were slowly migrating away from the burn areas and adversely affecting extended areas of the Shroud. Based on the proposals made in 2000, a special project was discreetly approved in November 2001 by Pope John Paul II to do the necessary work to "preserve and protect" the Shroud. Several interventions were made during the subsequent June 2002 project, including the following: 1. The original patches that covered the burn holes from the 1532 fire were removed. Loose debris and char associated with the burn holes was removed. Char around the much older "L" shaped poker holes was also removed (see Item L8). 2. The Shroud was turned over, and the Holland backing cloth that had covered the back of the Shroud for more than 450 years was unstitched and removed. 3. The back of the Shroud was lightly vacuumed to remove char and "debris" that had accumulated between the Shroud linen cloth and the Holland backing cloth. Some "debris" was also cleaned from the front of the Shroud. Unfortunately, some of this "debris" was potentially important archaeological evidence. 4. Following the work of removing the Holland backing cloth and "cleaning", new spectrophotometry and digital scanning of both the front and back of the Shroud was performed. Additionally, new high-definition photographs of the front and back of the Shroud were taken. 5. A new linen backing cloth was stitched to the Shroud. Each burn hole was also reinforced with surrounding stitching using curved needles and low-tension-inducing nylon thread. Removed char and other "debris" was catalogued according to each area from which it was collected and preserved in glass vials. The 2002 interventions drew severe and justified criticism from various quarters. [2][3][4][5][6] Unfortunately, much of the planning for the preservation project was conducted without broad consultation. Many of the broad-spectrum of scientific, historical and archaeological disciplines that study the Shroud for the benefit of mankind were not asked about reservations they might have had with the planned work. Without doubt some important aspects of the "archaeological site" that is the Shroud of Turin were adversely disturbed by the preservation project. For example, in the process of removing char from the 1532 fire, some evidence of where old fold artifacts intersected the char area was lost. Another example is the scraping off of what was thought to be random debris. Such debris might have forensic significance. In the future, it is vitally important for the custodians of the Shroud to formally recognize that the Shroud is, indeed, an **archaeological site** containing

important scientific information that must be preserved for future generations. This fact, it is suggested, should impress upon the custodians of the Shroud that broad multidisciplinary oversight is merited.

H28 Today

Since the in-depth scientific study of the Shroud by STURP in 1978, the Shroud has continued to be studied by scientists throughout the world. Every year numerous new scholarly and scientific papers and symposiums on the Shroud are hosted in different locations around the world. Nevertheless, although the Shroud contends for the claim of being the world's most important object, for most of mankind it remains virtually unexamined and unknown in any detail, except for its familiar name, "The Shroud of Turin".

Shroud Reliquary [1]

The word "reliquary" is used to refer to a container for a holy relic, an object believed to be part of a deceased holy person's body or a belonging held as an object of reverence. Today the Shroud is stored in a flat and horizontal position in a specially constructed high technology reliquary. The base of the reliquary is a single aluminum alloy casting with a milled recess for the Shroud of approximately 4 inches (10.2 cm). The top of the reliquary is made of thick tempered glass that is treated to protect the Shroud from ultraviolet light. When the Shroud is in the reliquary, the glass top is sealed hermetically to create an air and watertight environment. Once sealed, the reliquary is charged with inert gas (argon) and secured in a dark vault built into the left transept of the Turin Cathedral. The Shroud's environment is monitored continuously by a sophisticated system that measures temperature, pressure and humidity. When the Shroud is taken out of its normal storage reliquary for public expositions, it is encased in a second high-tech reliquary that allows for the Shroud to be displayed in a vertical position.

(Fig. 44) The modern high-tech Shroud Reliquary

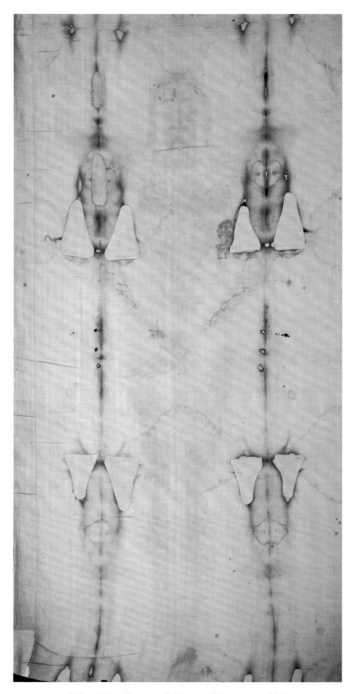

(Fig. 45)　Frontal Shroud image　(SI)

Section 2: Medical Forensics Evidence

IDR	Evidence/Comment
M1.1	The photographic negatives below show the frontal and dorsal body images. Bloodstains, which are dark on the actual Shroud, show as light or white in the negative images. [1][2][3][4] ← Blood around top of head, on forehead and in hair at sides of face ← Scourge wounds on chest ← Large blood flow from wound to right side of chest ← Blood flows on arms ← Blood flow at wrist that is consistent with a nail exit wound ← Scourge wounds down front of legs both above and below knees ← Blood flow at top of feet **(Fig. 46) Negative of frontal image (BI)**

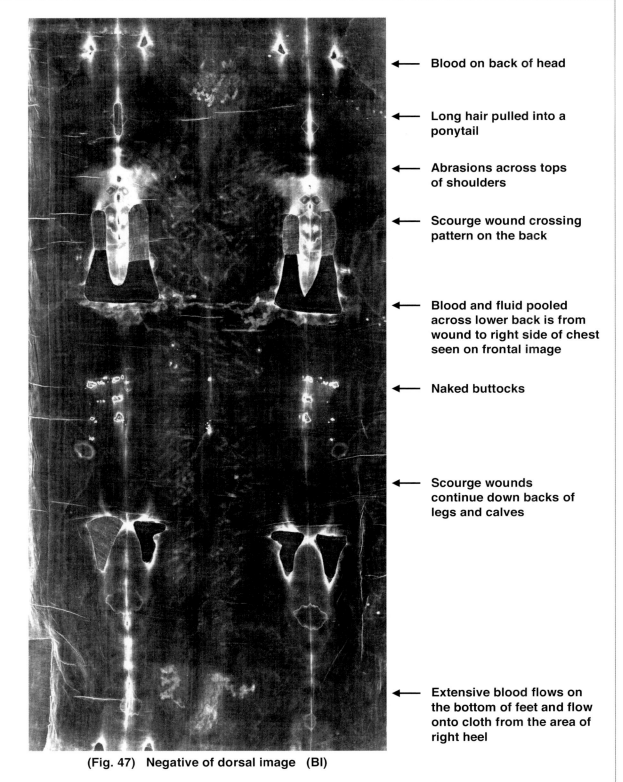

(Fig. 47) Negative of dorsal image (BI)

Medical doctors and forensic scientists have studied the body images on the Shroud for more than 100 years. Intense forensic evaluation of the Shroud began as soon as the photographic negatives of Secondo Pia were released to the public in 1898. The negative images show the wounds and blood flows in great detail.

M10 1	**The forensic evidence is consistent with the feet being nailed to the cross.** [1][2][3][4][5][6]

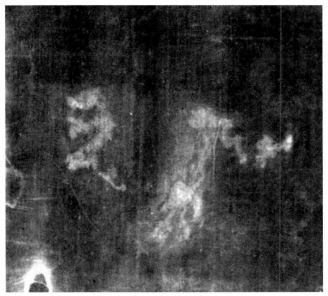

The details of precisely how the feet were nailed to the cross are open to interpretation. TSC's interpretation is that one foot was placed on top of the other foot, and then a single nail was driven through both feet into the crucifixion post (**Stipes**). Archaeological finds in Jerusalem have shown that in other cases the feet were turned to the side and nailed through the heels (see Item H23.) with a long single nail. In either case, the nailing of the feet would have been excruciatingly painful. The blood flows on the Shroud appear to be consistent with either method of nailing the feet.

(Fig. 57) Detail of blood on the bottom of the feet and blood flow onto cloth off of the right heel (BI)

M11 1	**The Shroud shows that gravity affected the blood flow associated with the wound to the left wrist. The alignment of the blood flow relative to the direction of gravity has been used forensically to show that the arms were raised at an angle of approximately 20 degrees from the horizontal while the man was suspended on the cross in the death position and that the shoulders were likely dislocated.** [1][2][3][4][5]

Hanging in this position would lead to extreme pain in the shoulders and arms. Also, the chest would be stretched and, thus, compressed. Breathing would be difficult. The remedy would be to push up on the impaled feet in order to gain some relief. One agony would be joined to another. Movement to gain any respite would cause exhaustion and profuse sweating. There would be no middle ground where the agony would subside.

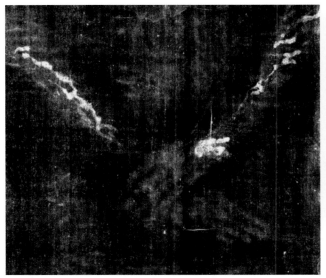

(Fig. 58) Close-up of hands (BI)

(Fig. 59) Arm at angle of crucifixion (BI)
↓
Direction of Gravity

	Fig. 58 above shows a close-up of the back of the left hand and forearm. Fig. 59 shows the left arm rotated to show the approximate angle from the horizontal of the arm when the body was in the crucifixion position. In this image, one can see how gravity might have affected the direction of the blood flows on the wrist and the forearm.
M12 3	**The exact cause of death of the Man of the Shroud is disputed.** [1][2][3][4][5] There is an image on the Shroud showing the wounds of the condemned, scourged and crucified victim in great detail that can be studied forensically, but there is no body available for a full autopsy. Nevertheless, the weight of the forensic work tends to favor a judgment that death resulted from a combination of hypovolemic and traumatic shock from the scourging and crucifixion. Hypovolemic shock, also known as hemorrhagic shock, results when you lose more than 20 percent of your body's blood or fluid supply from bleeding and severe sweating. If untreated, hypovolemic shock can result in the heart being unable to pump a sufficient amount of blood to the body; death follows.
M13 1	**The Shroud image shows a wound to the right side of the chest.** [1][2][3][4][5] 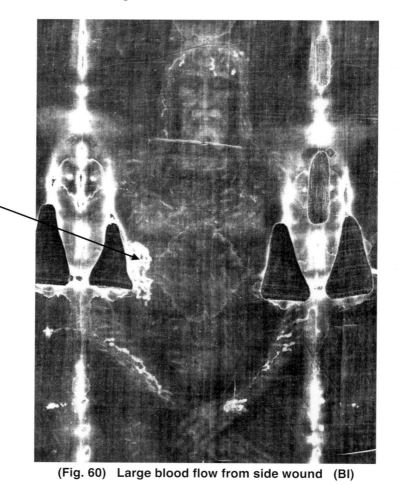 (Fig. 60) Large blood flow from side wound (BI) The image of the back of the body (Shroud dorsal image) also shows a large volume of blood and fluid from this wound pooling under the back of the body as it lay in the Shroud. The weight of forensic evidence shows that the wound to the right side of the chest was post-mortem. A post-mortem thrust of a lance into the thoracic cavity, delivered to insure the crucifixion victim was dead, is consistent with the release of blood and a massive pleural effusion of fluid from the area around the heart. This fluid would have accumulated because of the trauma of the brutal scourging coupled with the trauma of being crucified.

M14 1	**The Shroud image shows no evidence that the legs of the victim were broken.** [1][2]	
	The legs of a crucifixion victim were sometimes broken or significantly injured in order to cause rapid death. The breaking of the legs would remove the ability of the victim to distribute weight on the impaled feet and push up to aid breathing. Also, the fracturing blows would cause additional severe traumatic shock, and death would come quickly. If it were known the Shroud victim was already dead, then there would have been no need to break the legs.	
M15 1	**The image on the Shroud shows only the four fingers of each hand. The thumbs are folded under the palm.** [1][2][3]	
	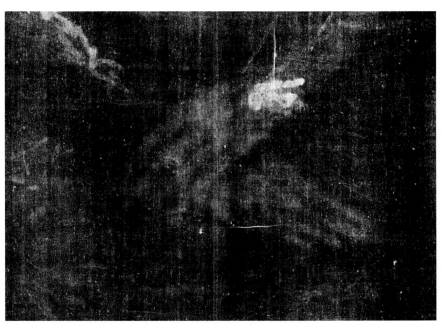	
	(Fig. 61) Close-up of Shroud hands (BI)	
	Some forensic scientists think that the traumatic puncture wounds of the crucifixion nails through the wrists would have damaged nerves, causing the thumbs to rotate in toward the palm.	
M16 1	**The blood on the Shroud has been shown to be human blood.** [1][2][3][4][5][6]	
	In 1978 during testing in Turin, the STURP team collected tape samples from the Shroud, including samples from alleged blood areas. **STURP** biophysicist John Heller and chemist Alan Adler studied the samples and collaborated in publishing a report in 1980 in which they confirmed the presence of actual blood on the Shroud. In 1981 Heller and Adler extended their research to include serological techniques that involved the diagnostic identification of two major blood serum proteins: albumin and immunoglobulin (antibody). They used the results of these tests to conclude that the blood is primate/human blood. (The bloodstains are discussed in more detail in Section 4 Item B6.)	
M17 2	**The body is in rigor mortis.** [1][2]	
	Most forensic scientists who have studied the Shroud image concur that the body appears to be in a state of rigor mortis. In particular, analysis of the dorsal body image supports the conclusion that the body depicted on the Shroud is rigid and in a state consistent with rigor mortis. This is easily observed, particularly in the area of the buttocks where there is no observed flattening due to body weight.	

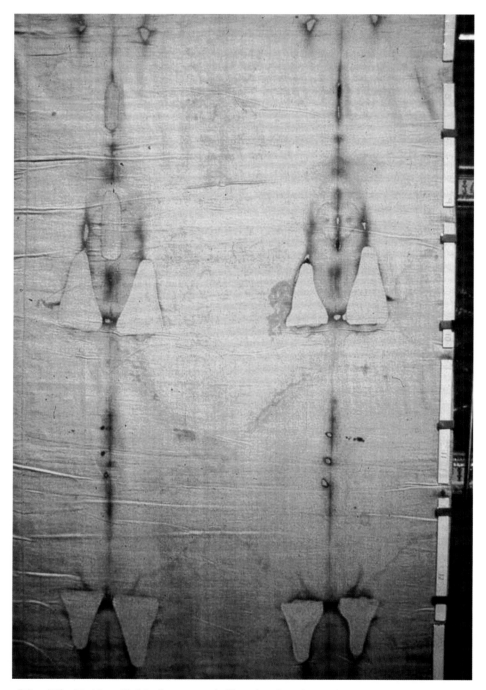

(Fig. 62) Raking-light photograph illuminating frontal image fold lines   (SI)

Section 3: Linen Cloth Evidence

IDR	Evidence/Comment
L1 1	The Shroud conservation project of 2002 stabilized the layout of the Shroud by stretching it out for flat storage. The reported post-preservation dimensions are 14' 6" x 3' 9" (4.42 x 1.14 m). Prior to the 2002 preservation the dimensions most often used for the Shroud were 14' 3" x 3' 7" (4. 34 x 1.09 m). The picture below, dating from the 1978 STURP expedition, shows the Shroud being viewed directly in natural light and illustrates the location of the frontal and dorsal body images. The images of the body are a mirror image of the actual body. [1][2] 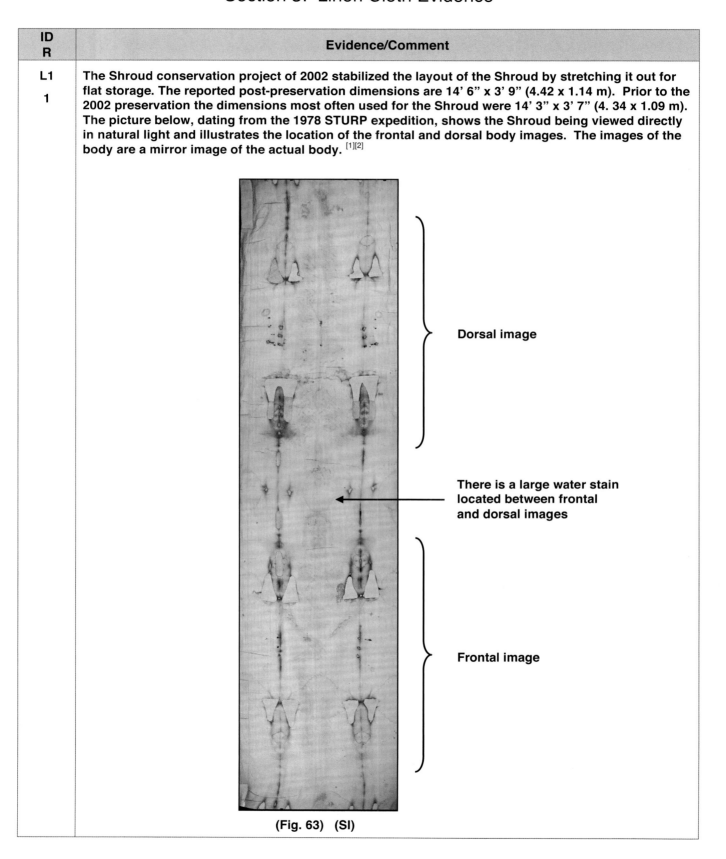 (Fig. 63) (SI)

		The Shroud was not woven to these particular specifications. Instead these dimensions are only approximate measurements for an ancient cloth that has been handled, stretched in varying ways and manipulated for centuries. Consequently, the more accurate specified dimensions of the Shroud, that is the dimensions used by those who crafted the Shroud, are more likely in cubits. A weaving specification of the Shroud of **8 cubits long x 2 cubits wide** would conform closely with the ancient Assyrian cubit of approximately 21.7 inches (55.1 cm) that was used in the area of Palestine in the first century. [3]
L2 1	**The linen Shroud cloth is nominally .014 inches (0.35 mm) thick, woven of threads of a mean diameter of .010 inches (0.25 mm). Each thread is composed of 70-120 linen fibers; each linen fiber is between .0004 inches (0.010 mm) and .0008 inches (0.020 mm) in diameter, which is less than the diameter of a typical human hair.** [1][2][3][4][5]	
	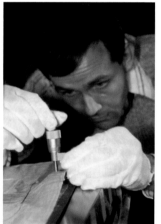 (Fig. 64)	Physicist John Jackson measured the cloth thickness at the time of the STURP expedition, using a micrometer that was zero checked. STURP colleague Ray Rogers recorded the measurements as Jackson systematically proceeded with the measurements. The measured thicknesses were in microns: 1. Frontal part of Shroud: 350, 342, 355 2. Dorsal part of Shroud: 391, 358, 348, 363 3. Dorsal foot area: 318, 313, 331 It should be noted that Jackson made no measurement below 300 microns. It should also be noted that due to the fact that the threads are handmade the number of fibers per thread is not uniform.
L3 1	**The cloth is woven in three-to-one herringbone twill. The picture below shows an area that is close to the feet of the Shroud image.** [1][2][3]	
	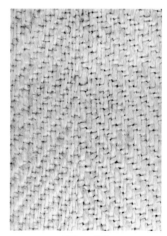 (Fig. 65)	**Warp:** These are the threads that are strung onto the loom before weaving begins, usually in a vertical direction. They run the length of the cloth corresponding to its long fourteen-foot plus dimension. **Weft:** These are the threads that run across the loom, being passed over and under to create the cloth. For the Shroud the weft or cross thread passes over three warp threads, under one, over three in a repeat pattern across the full width of the warp threads on the loom. Each succeeding weft thread is offset one warp thread either to the right or left. **Herringbone:** This simply means the offset or twilling is periodically reversed. The appearance is likened to a herring fish bone.
L4 1	**The weave and particular stitching are very distinctive and rare.** [1][2][3][4]	
	Nothing comparable to the Shroud has been found that originated in medieval Europe. The late John Tyrer, a textile researcher in Manchester England studied the X-radiographs of the Shroud and stated: *"the Shroud is a very poor product by comparison (to medieval European fabrics). It is full of warp and weft weaving defects. The impression I am left with is that the cloth is a much cruder and probably earlier fabric*	

than the backing and patches. This I think lifts the Shroud out of the Middle Ages more than anything I have seen about the textile." [5]

The radiocarbon dating of the Shroud was done in 1988 under the project management of the British Museum. Michael Tite, the lead manager on the project for the British Museum, conducted a thorough search for a control sample from the Middle Ages that would reasonably match the Shroud. "**He could find nothing.**" [6][7] On the other hand, archaeologists have discovered ancient wool artifacts with a herringbone weave similar to the Shroud. The artifacts were found in the ruins of a Roman fort in Egypt that dated from the 1st century. Mechthild Flury-Lemberg, the textile expert who was in charge of the 2002 Shroud preservation project in Turin, has said that even though the Shroud has many weaving defects, the herringbone weaving pattern itself would have been considered very special in antiquity in Palestine. [8]

| L5 1 | **Backlit photographs of the Shroud linen show darker and lighter banding in both the vertical and horizontal direction that corresponds to the warp and weft threads. There is more intensity of the banding corresponding to the weft threads, that is, across the Shroud in its shorter dimension.** [1] |

The banding is difficult to see in normal light. Some banding can be seen in the positive photographs and more can be seen in the negative images. However, when contrast is computer enhanced, the vertical and horizontal banding is easily discerned, particularly in backlit photographs. Banding can result when individual collections of flax have a slightly different color as a result of the collections being retted and bleached separately. Retting is the process of soaking the flax in water to separate the linen fibers from the main stalk of the flax plant. Separate batches of flax are then woven into hanks of thread and mildly bleached. This observation of bands of color conform to **Pliny the Elder's** (23AD – 79 AD) documented method of producing ancient linen. [2] Medieval linen was manufactured differently, and surviving high quality medieval linens (none found with a herringbone weave) do not show banding such as that found on the Shroud.

| L6 1 | **Raking or grazing light photographs of the Shroud show old fold marks in the linen cloth.** [1][2][3][4][5] |

One of the tasks undertaken by the STURP team was to take raking light photographs of the Shroud. Linen has poor elasticity, explaining why it wrinkles so easily. Thus, linen cloth has sort of a memory that can reveal how the cloth has historically been folded (see Item H14,2). Some of the fold lines found on the Shroud are as sharp as a straight edge, and there is an intriguing discoloration band associated with one set of especially closely-spaced folds; as if these particular folds might be associated with the Shroud being folded over the edge of a wooden block or batten. The TSC research team developed a computer program that analytically mapped the prominent folds found on the Shroud and found the folds to be consistent with the design of a lifting device that could have been used for raising the cloth. [6]

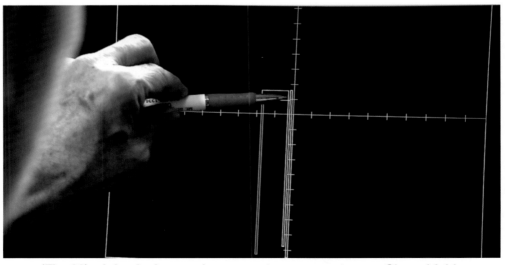

(Fig. 66) John Jackson using computer program to map Shroud folds

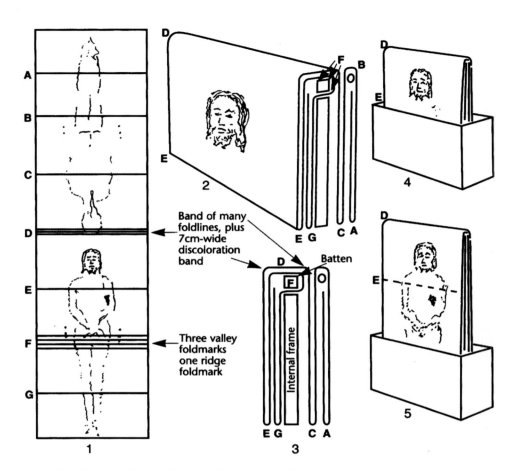

(Fig. 67) Schematic of Shroud lifting device based on computer fold analysis

(Fig. 68) Working prototype of Shroud lifting device

L7 1	A seam runs the full length of the Shroud approximately 3" (7.6 cm) from one edge. [1][2][3]
	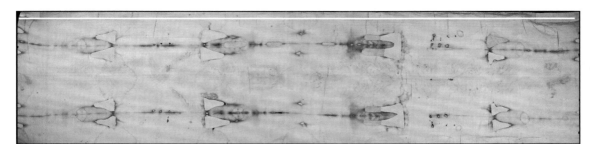
	(Fig. 69) Dashed line indicates location of the seam (SI)
	The purpose of the seam is uncertain. Some Shroud researchers think the side strip was cut from the Shroud and used to wrap the dead body that was once enshrouded in the cloth, and then at some later date re-sown back onto the main body of the Shroud. The stitching that reconnects the side strip to the main body of the Shroud is of note. Mechthild Flury-Lemberg, the textile expert who was in charge of the 2002 Shroud preservation project in Turin, found that the stitching pattern of the seam is similar to stitching found in the hem of a cloth that was discovered in the tombs of the Jewish fortress of Masada. The Masada cloth has been convincingly dated to between 40 B.C. and 73 A.D.
L8 1	There are four (4) matched repetitions of "L" shaped holes, generally referred to as the "poker holes." The "poker holes" predate the 1532 fire. [1][2][3]
	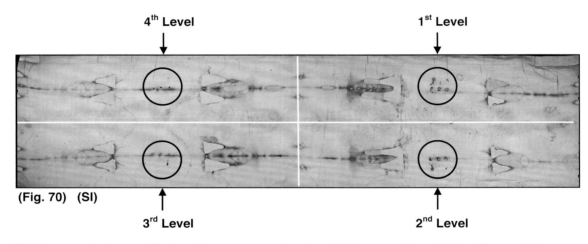
	(Fig. 70) (SI)
	The four matched sets of holes show a progressive level of damage. **Level 1** has the most damage and **Level 4** has the least. This pattern has been used forensically to determine that the Shroud was folded in four, first widthwise and then lengthwise. At one time, it was thought that a hot poker might have been thrust through the cloth. Today the favored hypothesis for the "poker hole" damage pattern is that hot incense fell on the cloth during its early pre-European history while the Shroud was being used in an ecclesiastical rite. The holes have also been linked to the Pray Manuscript (see Item H14, 5).
	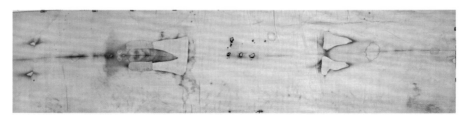
	(Fig. 71) Enlarged image of "1st Level" poker holes shown above. Note that the poker holes are located very close to the center of folded cloth.

| L9 1 | **There are large areas of burn damage, scorch lines and water stains on the Shroud, all associated with the 1532 fire in Chambéry, France where the Shroud was then kept (see Item H20). The eight (8) major burn blemishes on the Shroud are shown below. The geometrical pattern of the burns occurred when molten metal from the lid of the Shroud's storage casket fell onto the folded linen cloth inside the casket.** [1][2] |

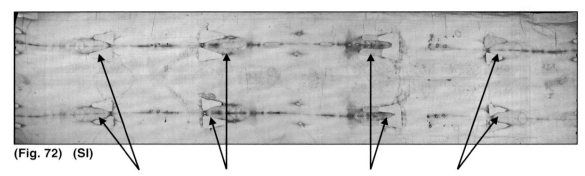

(Fig. 72) (SI)

Series of burn blemishes from 1532 fire

There are numerous water stains on the Shroud from the water used to douse the fire. These water stains are easily identified and are well documented. The folding pattern for this burn damage is very different from that associated with the "poker holes" discussed above in Item L8. In this case, the Shroud was folded into eight layers.

| L10 1 | **In addition to the water stains associated with the 1532 fire, there are many large water stains along the edges of the long dimension of the Shroud and along its central axis that are considered to be much older than the 1532 water stains, and potentially ancient.** [1][2] |

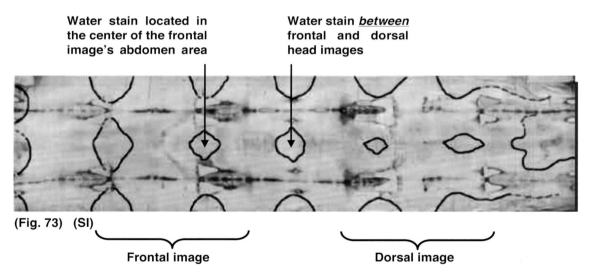

(Fig. 73) (SI)

Frontal image   Dorsal image

This Image is courtesy of Also Guerreschi and is slightly contrast enhanced

These water stains originated from an event or situation that predated the water stains associated with the **Chambéry** fire of 1532 and are potentially **ancient**. Two important papers by Aldo Guerreschi and Michele Saicito, published in 2002, show how the Shroud folding pattern associated with the large water stains is consistent with the Shroud being stored in an ancient type of earthenware jar at some point in its history before 1532. The hypothesis is that at some point water leaked into the vessel and wicked up into a corner of the folded Shroud through capillary action. As pointed out in the History Section (see Item H15), *The Man of Sorrows Icon* appears to incorporate the water stain at the center of the Frontal Image.

SECTION 3

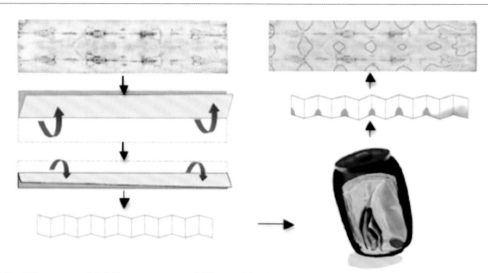

(Fig. 74) Diagram of folding pattern of Shroud before being placed in hypothetical storage jar

L11 2	During the time of the 1978 STURP expedition to Turin, sticky tape samples were taken that captured "dirt" from the Shroud. Subsequent studies of the "dirt" particles established that they were "quite similar" chemically to soil and stone typical of the area of Jerusalem. [1][2][3]

STURP team members Roger and Marty Gilbert, while doing ultraviolet spectroscopy scanning of the dorsal image area of the Shroud, detected unusual signals when they reached the area corresponding to the **soles of the feet**. They immediately called for the assistance of optical physicist Sam Pellicori who brought his portable Swiss-made Wild Heerbrugg M400 Microscope to examine the area. Shroud historian Ian Wilson has written that Pellicori's surprise statement was simply, *"It's dirt."* [4] Elevated levels of "dirt" were also found in correspondence to the areas of the **nose** and **left knee**.

1. **Kohlbeck and Nitkowski Studies:** [5][6][7][8][9][10] In 1982 STURP scientist Ray Rogers gave sticky tape samples to Dr. Joseph Kohlbeck, a crystallography scientist at **Hercules Aerospace** in Utah, for his analysis and comments. Dr. Kohlbeck found calcium carbonate (limestone) particles on the tapes from the foot area. This interested him because he knew that crystals of limestone can often give a "signature" that can point to a geographical source location. Kohlbeck was able to consult with a local Utah archaeologist, Dr. Eugenia Nitkowski (former Carmelite nun Sister Damian of the Cross), who had conducted archaeological research on Israel's rolling stone tombs. Nitkowski provided Kohlbeck with samples of Jerusalem limestone. Kohlbeck's subsequent analysis showed that the Shroud limestone sample closely matched the Jerusalem tomb limestone. Both appeared to be of the rare **Travertine Aragonite** variety of limestone found in the area of Jerusalem and only in a few other places on earth. To confirm his tests Kohlbeck took both samples to Dr. Ricardo Levi-Setti of the University of Chicago's Enrico Fermi Institute. Levi-Setti was the inventor of the high-resolution scanning ion microprobe, and the equipment at the Fermi Institute was state-of-the-art. Levi-Setti was able to compare the wavelengths emitted by the two crystalline samples. He confirmed that there was an unusually close match. Kohlbeck and Levi-Setti acknowledged at the time that they had not proven that the aragonite crystal found on the Shroud had come from a Jerusalem tomb. Their evidence was instead a pointer that must be considered and its weight judged in the context of other evidence.

2. **University of Padua Studies:** [11][12][13] During the 1978 STURP investigation and again in 1988 during the carbon dating project, Giovanni Riggi Numana, a colleague of Turin's scientific coordinator Professor Luigi Gonella, was authorized to use a mini-vacuum to collect dust samples from between the backside of the Shroud and the Shroud backing cloth. Between 2009 and 2011 the University of Padua (Italy) conducted a research project on Shroud dust samples that had been provided to them by Riggi. The Padua project was funded under the title *"Analysis of Microparticles Vacuumed from the Turin Shroud"*. The results were published in a technical paper issued by the Padua research team and were also reported in Giulio Fanti's important book, *"The Shroud of Turin: First Century after*

Christ!" The Padua research team reported that the particles taken from Mt. Zion (Jerusalem) were "quite similar" to the Shroud dust samples. The Padua research team stated in their paper that the soil particles analyzed were typical of Jerusalem but also of other arid "Mediterranean areas influenced by the winds of the Sahara Desert".

3. **Gérard Lucotte of the *Institute of Molecular Anthropology, Paris* Study:** [14] Lucotte received a tape sample from the nose area of the Shroud in 2005 from Giovanni Riggi Numana. Numana reported he took the sample from the nose area of the Shroud during the 1978 STURP expedition. Lucotte studied dirt particles found on the sticky tape by scanning electron microscopy (SEM) and by X-ray microfluorescence (XRMF). In 2015 Lucotte published results that "indicate a soil nature corresponding to desertic or semi-desertic climates."

The evidence associated with the "dirt" on the Shroud is compelling. The dirt found in the area of the dorsal foot area, in particular, is consistent with the concept of dirt being transferred to the Shroud from the feet of a barefoot man. The dirt on the nose and left knee would appear to be consistent with a fall or multiple falls to the ground. Also, the tests, performed independently on "dirt" found on the Sudarium of Oviedo that showed chemical signatures that closely correlated with dirt samples from the Calvary site in Jerusalem, can intriguingly be judged to add weight to the above reported results found for the Shroud "dirt" (see Item H11). It has become apparent that this area of research is **extremely important**. But its importance only became apparent after the STURP project. None of the STURP "dirt" samples taken were associated with a pre-planned protocol for testing that might determine a geographical correlation. Consequently, the sampling was limited, and the rigor in the custodial management of the samples that were taken has been justly criticized. Thus, TSC must reluctantly rate the "dirt" evidence as **Class 2** evidence; still significant, but the strength of the "dirt" evidence can and must be strengthened. The Shroud custodians ought to authorize new testing of dirt samples from the Shroud. It would be a straightforward project to solicit proposals for new research designed around specific testing protocols.

L12 / 2

The Shroud's "pollen fingerprint" is consistent with the Shroud being in the environs of Palestine, and more specifically, Jerusalem during its pre-European history. [1][2][3][4]

Some background on the evidential power of pollen is important. [5][6][7][8] There are roughly 380,000 species of plants that have so far been identified on earth. The scientific study of the pollen of these plants is the branch of the science of botany known as **Palynology**. In the 1970s and 80s a powerful new sub-branch of palynology emerged, known as **Forensic Palynology**. This discipline focuses on pollen found on an object of investigative interest (criminal, historic or archaeological) as potential evidence that can place that object in a certain place at a certain time of year.

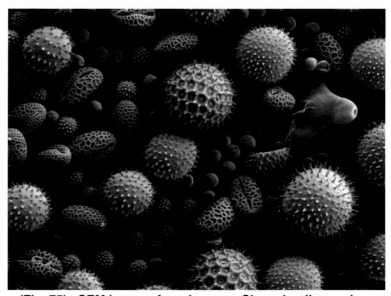

(Fig. 75) SEM image of random non-Shroud pollen grains

The foundation of Forensic Palynology rests on four remarkable scientific facts:

First: Pollen grains are extremely small, between 7-200 micrometers, so small that pollen grains cannot generally be individually seen by the naked eye. Yet pollen grains have very complex shapes and structures allowing many to be identified to the species or genus level. Because of their extremely small size, the use of a **Scanning Electron Microscope (SEM)** is generally required to be able to firmly differentiate individual pollen grains to the species or genus level.

Second: Pollen is extremely resistant to decomposition. The outer wall of a pollen grain, the "exine," is composed of **sporopollenin**, one of the most chemically inert biological polymers known. This polymer makes it possible for pollen grains to be preserved for millions of years.

Third: The surfaces of pollen grains are generally covered with waxes and proteins that are held in place by complex structures known as "**sculpture elements**" that enable pollen grains to stick or adhere to almost any "host", such as a linen cloth.

Fourth: Each geographical region on earth has its own unique pollen spectrum generated from the biodiversity of plants that grow there. This unique pollen spectrum for a specific geographical region represents a relatively indestructible **fingerprint** of the region.

1. **The Work of Max Frei (b 1913 – d 1983):** [9] Frei was recognized during his lifetime as one of the foremost criminal forensic scientists in Europe. In 1948, he was the founder of the Zurich, Switzerland **Central Police Scientific Department.** As the director of the department, he oversaw all the forensic science work, and under his direction the laboratory performed early pioneering work in **Forensic Palynology**.

(Fig. 76) Max Frei takes sticky tape samples during the 1978 STURP expedition while STURP chemist Ray Rogers of the Los Alamos National Laboratory looks on.

Shortly after his retirement in 1973 Frei's prominence led to an invitation from the custodians of the Shroud in Turin to join a small group of scientists for a brief and secret examination of the Shroud. That examination took place on 24 November 1973. Frei collected twelve (12) sticky tape samples from the surface of the Shroud for the purpose of collecting possible traces of pollen, and indeed, Frei's sticky tape samples did pick up traces of pollen. At the Central Police Scientific Department Frei had been the director of a staff that included professional palynologists; however, Frei had never been a practicing professional palynologist himself, although the thesis for his doctorate in botany had been on the subject of palynology. Nevertheless, Frei personally tackled the effort to identify the pollen grains on the tape samples he obtained from the Shroud. Over the next five years he made seven trips to various locations in Palestine, Turkey, France and Italy, collecting over 300 different regional pollen samples for comparison with the pollen specimens he had collected from the Shroud. In 1978 Frei was again invited by the Shroud custodians in Turin to take sticky tape samples at the same time STURP scientists were undertaking their own scientific study of the Shroud. He was able to collect twenty-six (26) additional tape samples from the Shroud. It is significant that Frei never extracted any pollen grains from these new 1978 tapes [10]; instead, he only used the new tapes to cross-check his continuing detailed work on the pollens extracted from his 1973 tapes. This cross-checking was done with relatively low-powered optical or light microscopic instruments to view pollen that was still adhering to the tapes. In June of 1982 Frei published an article in the special interest journal *Shroud Spectrum International*. [11] In this preliminary article Frei reported that in his several years of effort he had been able to identify the pollen of 56 plant species: some from Palestine; some from the region of Anatolia in Turkey; at least four from the region of Constantinople; and some from France and Italy. Frei's preliminary conclusion was that: "**The pollen-spectrum as described are a most valuable confirmation of the theory that the Shroud traveled from Palestine through Anatolia (Turkey) to Constantinople, France and Italy**." [12] Tragically, Frei died suddenly in January 1983 without leaving in place a team to carry his research to completion, to have it peer reviewed, or to have it formally published.

2. **The Supporting Work of Other Researchers:** [13] [14] [15] [16] The University of Padua research team that studied the dust samples vacuumed from between the Shroud and the Shroud backing cloth (see Item L11-2) reported finding pollen in the samples, and more specifically: "**a pollen grain of *Phillyrea angustifolia*, an evergreen plant that flowers between March and May and adapts well to the difficult terrain of some Mediterranean areas that are characterized by extreme drought. This type of pollen was just the type classified by Frei in his work.**" *Phillyrea angustifolia* was among Frei's list of 56 identified plants. Similarly, the study by the French scientist Gérard Lucotte of a tape from the nose area of the Shroud (see Item L11-3) also led to his discovery of pollen. Lucotte reported that he believed he was able to identify pollen, using SEM analysis, from two species of plants, ***Ceratonia siliqua*** (the carob tree) and ***Balanites aegypiaca*** (the Judas tree). Neither of these species was listed among the 56 reportedly found by Frei, but the two species of trees are commonly found in the region of Palestine.

(Fig. 77) Scanning Electron Microscope (SEM) photograph of pollen grain from the Shroud identified as *Phyllirea angustifolia*

		At the time Frei began his work, there were no well-developed databases of modern or fossil reference slides or fossil pollen assemblage slides of sediments of different ages for the areas of interest in Asia Minor, Palestine and Jerusalem. Also, there were few to no printed atlases for pollen from these areas and, furthermore, at the time there were no online internet pollen databases. Nevertheless, Frei's preliminary results, arguably buttressed by the later but limited pollen studies documented above, do offer evidence that the "fingerprint" of the environs of Jerusalem is on the Shroud. However, there are two significant reasons why this evidence must currently be rated **Class 2 evidence**. First, due to his untimely death, Frei never fully published his results, nor did he have it peer reviewed. Second, it is not clear that Frei used SEM-based analysis to support all of his reported findings. [17] Additional research is clearly required to upgrade the Shroud's forensic palynological "fingerprint" to **Class 1 evidence**. A clear path would appear to be open for such research. Today, unlike the 1970s and 80s when Frei did his work, detailed pollen atlases and online databases of pollen morphology are available for the areas of interest. Simply put, since the time of Frei the discipline and capabilities of professional Forensic Palynology have taken giant steps forward. The essentially untouched 1978 Frei samples mounted on slides were acquired in 1988 from Frei's widow by the United States-based organization known as ASSIST (Association of Scientists and Scholars International for the Shroud of Turin). [18] In 1993 the same Frei Collection was transferred from ASSIST to the personal custody of Shroud researchers Alan and Mary Whanger. [19] There is every reason to believe that the 1978 slides remain a palynological treasure chest that can be the focus of further studies. Even more important may be the well-catalogued collection of aspirated dust and dirt from the Shroud that is in the possession of the Shroud custodians in Turin. This material is clearly a resource for further study, and possibly, new studies could also be coordinated with a new **Forensic Palynology** study of pollen on the Sudarium of Oviedo (see Item H11). [20] [21]
L13 2	**Images of flowers have been found on the Shroud that demonstrate the Shroud was in the region of Jerusalem at some point in the past.** [1][2]	
	Before the Frei Collection was acquired in 1988 by ASSIST (see Item L12), Frei's widow loaned four (4) of the 1978 slides to research archaeologist Paul Maloney, who served as the General Projects Director for ASSIST. During his careful examination of the four slides, Maloney noticed debris that looked to him like the anther of a flower. He consulted with the palynologist, Dr. A. Orville Dahl of the University of Pennsylvania, who examined the slide in question and confirmed the presence of a plant anther, and further stated that he could count at least 11 pollen grains still inside the anther. [3]	

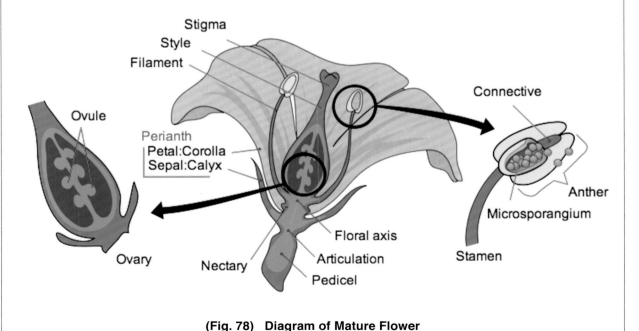

(Fig. 78) Diagram of Mature Flower

Later, after ASSIST received Dr. Frei's unpublished work and the full collection of 1978 tapes and slides, Dahl was given access to study all of the material. Dahl noted that of the 56 plant pollens Frei reportedly found on the Shroud, a majority of 32 were **entomophilous** (insect carried). This type of pollen is generally associated with flowering plants while **anemophilous** (wind carried) pollens are generally associated with non-flowering plants, e.g., pine trees. This fact, coupled with Dahl's viewing of the anther of a flowering plant on one of Frei's slides, led Dahl to propose that flowering plants at some time in the past had been physically laid on the Shroud. Alan Whanger, also an ASSIST researcher, independently reported that during his extensive study of Shroud photographs he had detected what appeared to him to be images of flowers. [4] One day in 1995 the Whangers visited the home of the prominent Jewish botanist, Avinoam Danin (b. 1939 – d. 2015). At the time Danin was broadly recognized as the world's foremost authority on the flora of Israel, particularly the region surrounding Jerusalem. The Whangers showed Danin photographs of the Shroud and asked him if he saw any images of flowers. Danin reported that he "**looked for some ten seconds and said that I saw images of a few plants I know from the Jerusalem area.**" [5] From this point forward to the end of his life, Danin became deeply involved in the study of botany related to the Shroud. Following many years of research, he published his results claiming to have identified images of seventeen (17) different plant species on the Shroud. He stated that his research convinced him that "**the origin of the Shroud is from an area between Jerusalem and Hebron; only in that area could people bring fresh plants of these species from the field and put them onto a dead man's body. These plants indicated that the time of year was March through April.**" [6]

TSC has no doubt that plant debris and images of flowers are on the Shroud. We can confirm that TSC staff members have reported seeing flower images, some conforming to those listed by Danin, on unenhanced full-size high-definition Shroud photographs at our Shroud Center in Colorado. However, we cannot make out many of the flowers Danin describes in his research. The perception of most flower images ultimately rests on the analysis of photographic images of the Shroud using computer software for image manipulation and enhancement, and even then, some "coaching" is generally required to "see" all of Danin's flower images. In a paper commenting on flower images on the Shroud, as well as other images such as coins and lettering reported by some to be "seen" on the Shroud, Shroud researchers Murra and Di Lazzaro wrote: "**Interpretations of shapes, coins, faces, flowers or letters "seen" on acheiropoieta images by means of image processing tools should be considered a track useful to address further studies, but they cannot be considered as self-consistent proofs.**" [7] TSC is in general agreement with Murra and Di Lazzaro's position [8], and thus, must rate the flower-image evidence said to locate the Shroud in the vicinity of Jerusalem at some point in the past as strong, but currently only as circumstantial **Class 2 evidence**.

L14	**Images of coins and symbols have been found on the Shroud.** [1]
3	The identification of these reported images on the Shroud must today be rated as **Class 3 evidence**. It is possible that enhanced image processing techniques may in the future strengthen the evidence for such images on the Shroud,

Section 4: Image Characteristic Evidence

Modern scientific research has revealed scores of unique features of the body and blood images on the Shroud. [1][2] In this section just seventeen (17) **Class 1 Evidence** image characteristics are listed. These seventeen image characteristics are judged to be the most important because collectively they are sufficient to critically evaluate all of the major image-formation hypotheses that have been proposed to date. This section of the Critical Summary is arguably the most important and should, therefore, be studied with the most care. You are encouraged to consult the references, as appropriate, to help you reach a better understanding of any specific image characteristic. The seventeen characteristics are logically broken down into two categories:

1. Image Characteristics Related to the Cloth.
2. Image Characteristics Related to the Body.

ID R	Image Characteristics Related to the Cloth
C1 1	**The frontal and dorsal body images have optical densities that are nearly the same. This means the relative lightness and darkness of the frontal and dorsal images are essentially the same.** [1][2]
	STURP determined that the maximum optical densities of the frontal and dorsal images are nearly the same. Because of this, it is difficult when viewing the Shroud to judge through normal human visual perception which image, frontal or dorsal, is darker. STURP researchers also confirmed that the Shroud image is continuously shaded to some degree over its full extent. In all areas of both the frontal and dorsal images, there is some discoloration of the fibrils of the threads, except at the location of the bloodstains.
C2 1	**The image is extremely superficial, with the image being present on only the very surface of the cloth. The colored linen fibers of the image lie only on the uppermost portions of the threads, leaving the inner fibers of the threads uncolored.** [1][2][3]  **(Fig. 79)** Image-bearing thread fibers in the area of the bridge of the nose

	This item addresses the remarkably superficial nature of the image. The linen Shroud cloth is nominally .014 inches (0.35 mm) thick, woven of threads of a mean diameter of 0.10 inches (0.25 mm), each of which is composed of 70-120 linen fibers, each in turn averaging between .0004 inches (0.010 mm) and .0008 inches (0.020 mm) in diameter. The diameter of a linen fiber, thus, is generally less than a typical human hair (see Item L2). Fig. 79 above shows a highly-magnified close-up of the threads and image-bearing linen fibers at the bridge of the nose. This is one of the most densely colored areas on the Shroud. Yet even here, the photograph gives a hint of just how superficial the image is. The full details of superficiality were revealed by STURP through high-powered microscopic examination (32X and 64X magnification) and in transmitted light photographs that allow comparison of the faint image to the much darker fire-related scorches found on the Shroud. At the thread level only the surface linen fibers bear the color of the image. 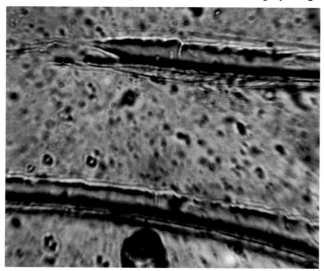 **(Fig. 80) Ray Rogers' photograph of "ghosts"** Some research suggests that the color of the image lies on the 0.2μm (micrometer or micron) thick layer interpreted as the primary cell wall of the fibers, with the cellulose of the medulla, the interior of the fibers, being colorless. [4][5][6][7][8] STURP research, along with other research, confirms that the depth of discoloration on individual colored image fibers is extremely shallow or thin. But interpretations vary, and definitive research has not yet confirmed that the color is restricted to the primary cell wall. STURP member Alan Adler reported that the thin, colored layer on image fibers could be reduced with a special chemical, known as a diimide reagent, leaving colorless, undamaged linen fibers behind. Other research demonstrated that the STURP tape samples had in many cases pulled off the colored layer on image fibers and left these "***ghosts***" attached to the sticky tape samples. Work remains to be done to identify the exact nature of these "ghosts".
C3 1	**The frontal image, at least in correspondence to the area of the face, is doubly superficial. This means that the .014 inch (0.35 mm) thick fabric presents a superficial image on the front of the cloth, no image in the middle of the cloth, and another superficial image on the backside of the cloth.** [1][2][3][4] This image characteristic refers to the opposite side of the cloth from the side normally associated with the colored fibers of the image. After the fire of 1532 Poor Clare Nuns added patches over the burn holes left by the fire and sewed on a support backing cloth that became known as the Holland backing cloth (see Item H20). At the time that these repairs were made, various bloodstains that had penetrated through the thickness of the Shroud were documented, but there was no mention of any image on the backside. In 2000, part of the Holland backing cloth was unstitched to allow for the passage of a scanner between it and the Shroud to facilitate a better examination of the backside of the cloth. In a 2004 paper entitled "***The double superficiality of the frontal image of the Turin Shroud,***" [5] G. Fanti and R. Maggiolo reported that their studies of the scanning photographs from the 2000 study of the backside of the Shroud showed that there is a very faint image of the face, hair, and possibly, of the hands on the backside of the cloth. Images on the backside of the cloth were in register with corresponding frontal body images.
C4 1	**There is no superficial image on the backside of the Shroud opposite to the dorsal image. Double superficiality artifacts (an image on both the front and back of the Shroud cloth) exist only corresponding to the frontal image.** [1][2] Fanti and Maggiolo's paper [3] states that their analysis of the photographs from the 2000 study of the backside of the Shroud did not reveal any image on the backside of the dorsal body image. Only double superficiality for some areas of the frontal image was discovered.

C5 1	**The fibers are only colored (yellow-brown) due to chemical reactions involving the polysaccharides composing the linen fibers: oxidation, dehydration and conjugation.** [1][2][3] The colored linen fibers are only colored due to a chemical reaction involving the fibers themselves. There is no evidence of a coating or extraneous material added to the fibers to cause the image color. The image-bearing fibers have a yellow-brown color. One of the primary goals of the STURP scientific expedition was to test the hypothesis that the Shroud image was painted. STURP testing results showed that no paint pigments or paint-carrying mediums could be found bound to image-bearing linen fibers. After the STURP work was completed in Turin, further testing was done on sticky tape samples with laser-microprobe Raman analysis, pyrolysis-mass-spectrometry and micro-chemical testing. In none of the testing was any evidence found to indicate the colored image-bearing linen fibers were coated with any paint pigments or bear any evidence of paint mediums or other extraneous matter. To say that the image-bearing fibers carry no paint pigment or paint medium, however, is not to say the Shroud itself does not carry any paint debris. During medieval times artists made copies of the Shroud, and many copies were subsequently laid on top of the Shroud to "authenticate" them as true copies. This practice would inevitably leave traces of paint fragments on the cloth. Also, iron oxide, a common compound found in medieval paint mediums, has been found on the Shroud. However, iron oxide appears to be evenly distributed over the entire cloth in both image and non-image areas, except in the bloodstains where it is highly concentrated, as would be expected. It is likely that the iron oxide came from soaking the flax in water as part of the retting process.
C6 1	**There are no signs of cementation between cloth fibers/threads or of capillary flow associated with any viscous paint or other artistic mediums being used to create the Shroud image. By contrast, the bloodstained areas show cementation and signs of capillary flow.** [1][2][3] 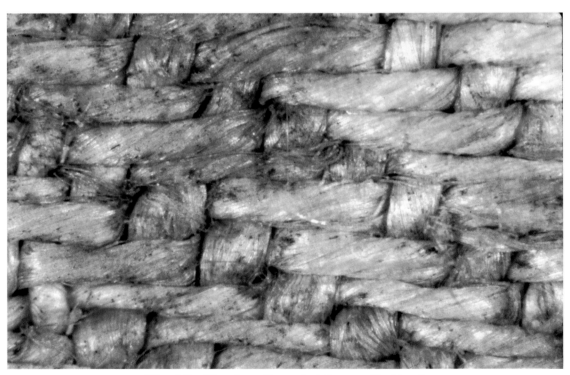 **(Fig. 81) Blood on Shroud at the small of back** In the area of the bloodstains and in some areas associated with water stains, there is clear and well-documented evidence of cementation between fibers, as well as evidence of capillary flow of liquid. In the areas where image fibers exist in isolation from such stains, there is no evidence of cementation between

fibers or evidence of capillary flow. Compare the image above of a bloodstain area with the image for colored fibers at the bridge of the nose, discussed in Item C2.

ID R	Image Characteristics Related to the Body
B1 1	**The Shroud images viewed directly in natural light have the tones of light and dark reversed with respect to what is normally experienced in human visual perception.** [1][2][3] 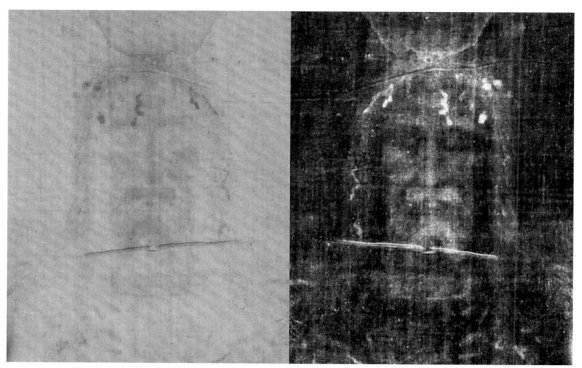 (Fig. 82) Photographic positive (SI)　　(Fig. 83) Photographic negative (BI) When viewing a negative photograph of the Shroud, the image details are dramatically easier to perceive. This difference of being able to perceive much greater detail when the light and dark areas of the Shroud images are reversed is the source of the so-called "negativity effect" of the Shroud image. As mentioned in the Historical Evidence section (see Item H22), Italian photographer Secondo Pia took the first official photographs of the Shroud in 1898. He had been invited to photograph the Shroud while it was being exhibited to the public in the Turin Cathedral. As he developed his film he was shocked to see that his negatives revealed previously unperceived details of an anatomically correct, naked and crucified man.
B2 1	**The image has a resolution at least as good as 1/5 inch (.5 cm) with no well-defined outlines or borders. This means that the image details such as the nose, lips, and beard are clearly defined, but the image on the Shroud seems to disappear if observed at a distance any closer than one meter.** [1][2] In an early STURP research paper it is stated: *"It is possible to estimate the apparent lower limit of resolution. Using the smallest anatomical feature discernible in microdensitometer scans of the image, probably the lips, we estimated that image resolution is at least as good as 0.5 cm."* [3] This level of resolution means that image features as small as 1/5 of an inch (.5 cm) can be clearly observed in negative photographs. It is also noted that when viewing the actual Shroud, the image essentially disappears when observed from closer than approximately 3 feet (≈1 meter), or inside an average arm's length away from the image. Jackson personally recalls that during the STURP expedition that, for him,

	image details could not be perceived when he was any closer than approximately six feet (1.83 meters) from the cloth. This inability to perceive details as one gets closer to the Shroud is due to the combination of the low contrast between the colored image fibers and the uncolored background fibers, and because there are no defined image borders. The ability to perceive details at closer range is remarkably and dramatically different with the photographic negatives that were not available until 1898.
B3 1	**The image-density distribution of both front and back images can be correlated to the distance between an object having the shape and contours of a human body and a cloth covering that body. This is why many state that the Shroud is a 3-dimensional image.** [1][2][3][4] 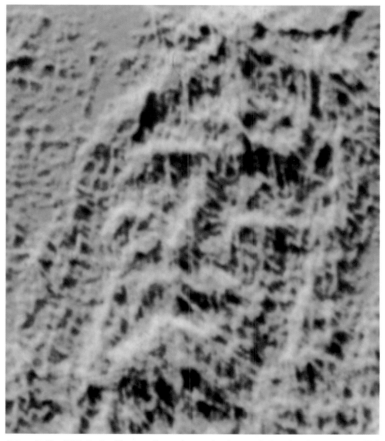 (Fig. 84) VP-8 3-dimensional rendering of the facial area (SI) The body image appears denser in the areas where the vertical distance of a body from an enveloping cloth surface would be shorter (closer = denser). This variation in the image density has been analyzed by both analog, and later, by digital computers to render 3-dimensional "maps" of the Shroud image, particularly of the frontal image. The 3-dimensional nature of the dorsal image is greatly attenuated. The back of the body was, for the most part, in direct contact with the cloth. Thus, the dorsal image on the Shroud more resembles a "contact" image, although there are some 3-dimensional attributes to the dorsal image as well. STURP worked to characterize the spatial distribution of the body image on the Shroud. It was demonstrated that there exists a relationship between *"the shading density of the image and expected cloth-body distances obtained by enfolding volunteer subjects in a full-scale model of the Shroud."* [5] That is, the closer the enfolding cloth is to the body, the denser the body image. For example, one of the densest body-image areas on the Shroud corresponds to the tip of the nose, a point where an enveloping cloth would likely be in direct contact with the body. There are two other ways of stating that the body image on the Shroud can be correlated with cloth to body distance. In some Shroud research papers, including the referenced Fanti paper, [1] these observations are

listed as separate items of evidence. We include these observations here as logical extensions of this one key item of evidence:

1. Body image fibers are visible in areas of non-contact zones between body and Shroud, for example, the nose and cheek areas.

2. The image density is less dense in non-contact areas than in contact zones. The convex "hills" of the face (e.g., the eyeballs and tip of the nose) are more clearly represented than the concave hollows; the tip of the nose is one of the most evident.

B4 1	**Image distortions of the hands, calves and torso are consistent with those that would be obtained if a body lying on its back were wrapped in the Shroud. The mapping of image features from the body to the cloth of the frontal image is more or less vertical, corresponding to the direction of gravity.** [1][2][3][4]

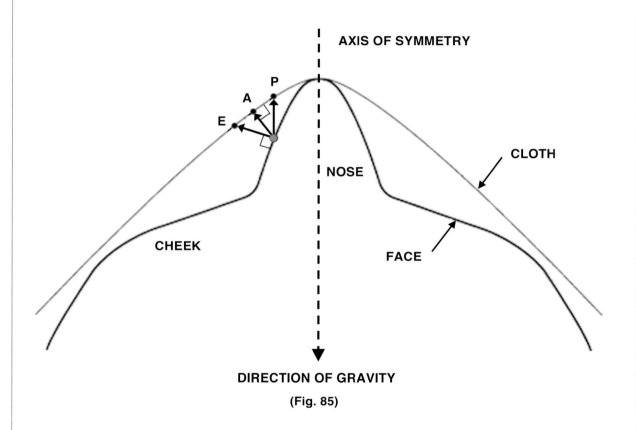

(Fig. 85)

Path E: **E**mission-dominated mapping.

Path A: **A**bsorption-dominated mapping.

Path P: **P**rojection-dominated mapping.

Consider the above diagram that shows a cross-section through the face (from ear to ear) at the level of the nose. This diagram illustrates the three-possible primary information transfer paths from a point (red) on the body to a corresponding point (black) on the cloth. **Path E** represents a path that is perpendicular to the body surface, **Path A** corresponds to a path that is perpendicular to the cloth surface, and **Path P** corresponds to a path that is vertical or parallel to the axis of body/cloth symmetry. This last path also corresponds to the direction of gravity if the body is lying horizontally on its back. Each of these mapping paths place the image of corresponding body points at a different location on the cloth. Thus, each mapping path yields different geometric "wrapping distortions" when the cloth is laid flat. Each of these three mapping

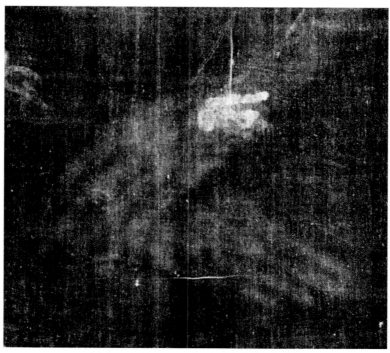

(Fig. 89) Close-up of the hands (BI)

Perhaps the metacarpal bones are easiest to observe in edge-enhanced photographs. The work of Dr. Alan Whanger and Mary Whanger has made a significant contribution in this area. The Whangers used a technique of image-edge enhancement for the hand images that shows the metacarpal bones quite clearly. [5][6]

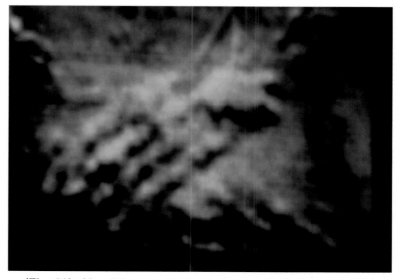

(Fig. 90) Alan Whanger "edge-enhanced" photograph (BI)

Similar techniques employed by the Whangers have suggested that facial bones and images of teeth can also be identified in the Shroud facial image. [7]

(Fig. 91) Colored image fibers in the area of the eyes

Compare the density of colored fibers in this photograph with the density of colored fibers in the area of the nose shown in Item C2.

Section 5: Image-Formation Hypotheses

Modern scientific study of the Shroud began in earnest shortly after Secondo Pia's negative photographic images became public in 1898. In the early years following Pia's release of his photographs, numerous image-formation hypotheses were proposed. Then, with the publication of the STURP research findings beginning in the early 1980's, the detailed empirical evidence concerning the image began to be universally understood. This empirical evidence allowed for a critical evaluation of older image-formation hypotheses and encouraged many new hypotheses. These hypotheses can generally be placed in one of three categories as follows:

1. **Dead Body Alone**: A process involving a dead body naturally causes the image.

2. **Human Artifact:** The image is the work of a human artist.

3. **Radiation/Electric Field:** The image is formed by radiation or in the presence of an electric field.

	Dead Body Alone
F1	**Contact Hypothesis (Vignon)** [1][2][3] The French scientist Paul Vignon (1865-1943) and his colleague at the Sorbonne Prof. Yves Delage were among the first generation of scholars to study the Shroud after the photographs of Secondo Pia became public in 1898. Vignon's first book on the Shroud was published in 1902. Thirty years later in 1933 Vignon was among a select group of scientists given permission to closely view the Shroud privately during an exhibition in Turin (24 September – 15 October, 1933). Unfortunately, this intimate encounter with the Shroud had to be undertaken without the availability of sophisticated instrumentation. Such an encounter with modern and sophisticated analytical tools would have to await the STURP project more than forty years later. In a 1937 article published by **Scientific American,** Vignon detailed the work and experimentation he and a team of researchers performed to investigate whether some natural process could produce the Shroud image. Vignon's team first considered whether a process involving contact with a dead human body that was wet with embalming oils or liquids associated with decomposition could be responsible for the Shroud images. Vignon concluded early on that the bloodstains were the result of contact with a dead human body, but the body images themselves could not be produced by contact alone. In the **Scientific American** article Vignon stated: *"After analyzing the first photographs of the Shroud and making our experiments in the laboratory of the Sorbonne, we concluded that the figures are the direct imprints of a human body. It was obvious at once that they were not produced by mere contact, for contact between the pliable cloth and the irregular surface of a human body would have caused considerable distortion, and there is little or no distortion in these figures."* [4] Others since the time of Vignon have considered and experimented with the possibility that the body images on the Shroud are the result of a natural contact mechanism with a wet dead human body. However, since the publication of the STURP research results that revealed the detailed nature and complexity of the Shroud image characteristics, all further efforts to promote a natural contact phenomenon for the image have been abandoned. There are simply too many inconsistencies and problems to overcome.
F2	**Gas Diffusion Hypothesis (Rogers)** [1][2][3][4][5][6] Paul Vignon also proposed a hypothesis involving gas diffusion of "**humid ammoniac vapors, resulting from the fermentation of urea, which is exceptionally abundant in the sweat produced by physical torture and by fever**". [7] He hypothesized that the "ammoniac vapors" reacted with aloes "**which were spread on**

the cloth and sensitized it to the action of the vapors". [8] It only took a single finding from the STURP research team to effectively rule out Vignon's vapor hypothesis. His hypothesis is not compatible with the Shroud image **superficiality** (see Item C2). Aloes spread on the cloth would penetrate the cloth, as would the "ammoniac vapors", meaning the image could not be restricted to only the topmost linen fibers on the surface of the Shroud.

Vignon's vapor-graphic hypothesis did inspire other hypotheses proposing a gas diffusion model. Most prominently, STURP chemist Ray Rogers made several refinements and extensions to the Vignon "vapor" theory. Rogers proposed a Maillard chemical reaction between amines that would hypothetically be generated by a decomposing body (putrescine and cadaverine that are both heavier than air) and an assumed microscopically-thin contamination layer on the Shroud fibers. He proposed that a contamination layer of starch was left on a very thin evaporation surface of the Shroud as a by-product of the manufacturing process of the ancient linen. Rogers demonstrated in experimentation with small samples of linen cloth that a Maillard reaction could lead (qualitatively) to yellow-brown coloring of a linen fabric that was treated to have a thin starch contamination layer. However, Heller and Adler in their paper "***A Chemical Investigation of the Shroud of Turin***" [9] could not identify any starch being present on the Shroud. Rogers reported that his experimentation spot tests with aqueous iodine indicated the presence of some starch factions on Shroud fibers. Nevertheless, even if the presence of a starch contamination layer on the Shroud could be demonstrated, there still appear to be many difficulties with the Shroud image being the product of a natural decomposition gas diffusion mechanism. Judged to be among the biggest inconsistencies with Rogers' gas diffusion hypothesis and with any other proposed gas diffusion mechanism, for that matter, are the problems with image **resolution** and the **orthographic** (vertical mapping) nature of the Shroud image (see Items B4 and B5).

Human Artifact

F3	**Painting Hypothesis (McCrone)** [1][2] The STURP scientists were authorized, after they had completed their on-site work in Turin (1978), to take home tape samples from the Shroud for further study. STURP scientist Ray Rogers transmitted a number of the sticky tape samples for analysis to respected microscopist Walter McCrone. McCrone subsequently reported that he found iron oxide on the samples, and from that finding he concluded that the iron oxide was evidence pointing to tempera paint, and therefore, to the painting of the Shroud image by a human artist. McCrone was correct in his finding of minute amounts of iron oxide on the Shroud tape samples. Shroud researcher Giulio Fanti has suggested that possibly some bloodstains were touched up with red paint during the middle ages to make them more dramatic for display purposes. [3] McCrone may have seen evidence for this as well. [4] However, McCrone may not have known that **STURP** scientists Morris and Schwalbe found that iron oxide was found to be distributed over the entire cloth, not just in the image or bloodstain areas. [5] The retting process, used to separate flax fibers from the rest of the flax stalk before they can be used to spin into linen thread, involves soaking the flax in water. From the first century to medieval times, pond retting seems to have been the preferred method. This process can lead to deposits of iron oxide being left on the flax fibers. Besides iron oxide, McCrone also reported finding actual paint debris on the tape samples he examined. This also was not an unexpected finding. Numerous painted copies of the Shroud were made in the Renaissance Period, and it has been historically documented that painted copies of the Shroud were "sanctified" by being laid on top of the original. Such sanctification would have inevitably left some paint debris. Nevertheless, McCrone persisted in his painting hypothesis. Ultimately, he wrote a full-length book supporting his hypothesis that the Shroud image was a human-created painting. Skeptics of Shroud authenticity continue to quote from McCrone's book. [6] In fact, McCrone's was the last of a long list of "painting" hypotheses that have all been discredited because all painting hypotheses are judged to be inconsistent with multiple image characteristics.
F4	**Dusting Hypothesis (Craig)** [1][2][3][4] Artists Emily A. Craig and Randall R. Bresee obtained one of the best results, from a macroscopic point of view, of producing a face image with many of the characteristics of the Shroud image. They used powdered

pigments to "paint" an image on paper, and then transferred the image to a linen cloth using a wooden burnishing instrument. The image was then "fixed" on the cloth with the aid of heat. Craig and Bresee limited their effort to creating only a facial image. They did not include any bloodstains with their image.

F5	**Bas-Relief Hypothesis (Delfino)** [1][2][3][4]

The term "bas-relief" is a French term that comes from the Italian "*basso-relievo*," which literally means "*low relief*". It is an artistic term referring to a sculpture technique. One variation on the bas-relief method was proposed and tested by the Italian, Pesce Delfino. Delfino used a metal sculpture of a human face, heated it to approximately 200° C (about 392° F), and then impressed a piece of linen cloth onto the sculpture in order to leave a scorched image. Delfino limited his effort to creating a facial image only. He also did not attempt to include any bloodstains or details of blood exudates from wounds.

F6	**Combination Human Body and Bas-Relief Frottage Hypothesis (Garlaschelli)** [1][2][3][4][5][6]

In 2008 and 2009 Luigi Garlaschelli, a researcher in organic chemistry at the University of Pavia in Italy, led a well-funded team effort to demonstrate how the Shroud image could have originated from the work of a medieval artist. On October 5, 2009 Reuters news service reported that Garlaschelli claimed his results "***prove definitively***" [7][8] that the linen cloth some Christians revere as Jesus Christ's burial cloth is a medieval fake.

The work of Garlaschelli's team represents the most extensive experimentally-based effort undertaken to date to reproduce a full "frontal" and "dorsal" Shroud image. The effort began with a rigorous study of the research on the nature of the Shroud image, along with careful consideration of all previously proposed natural (contact and gaseous), as well as artistic image-formation mechanisms. After rejecting hypotheses that proposed a natural image-formation process and other proposed artistic methods, the team settled on a **frottage** (French for "rubbing") technique as the most likely method "cunningly" used to paint the Shroud image (reminiscent of the words of Pierre d'Arcis in 1389). The effort produced impressive results that included a pseudo negative image that is fuzzy, resides on the topmost fibers of the cloth, has some 3-dimensional properties, and produces an image that does not fluoresce in the way that a scorch normally would. The Garlaschelli effort evolved through experimental trial and error to finally settle on a five-step process, as follows:

1. A pigment containing acidic compounds in a water-based slurry was rubbed on a prepared linen cloth molded over an **actual human model**. The human model was used for the frontal and dorsal images below the neck level. The acid used by Garlaschelli was sulfuric acid.

2. A shallow **bas-relief** was constructed out of plaster-of-Paris for the facial image and the same rubbing, or frottage method, was used to obtain the raw image. Garlaschelli reported that although a "Shroud-like image can be produced by a rubbing technique on a human body", the face "must be obtained from a bas-relief to avoid the inescapable wrap-around distortion".

3. Next the scourge marks were added. As Garlaschelli comments, "*...scourge marks and blood stains on the Shroud are not fuzzy but rather sharp. Thus, for our reproduction they were not added by rubbing. Instead, the pigment (this time a very diluted suspension of red ochre, cinnabar and alizarin in water) was gently applied with a small brush, which also gives rise to the fine, well-defined parallel 'scratches' seen in some of these marks.*" [9]

4. The cloth was then artificially aged. It was heated in a specially designed oven, then washed, and finally, ironed flat. This final heating process simulated what Garlaschelli proposed happened naturally to the "cunningly painted" Shroud over years and even centuries – the effects of time. That is, the Shroud image we see today is not at all the Shroud image that was first artistically created. The pigment essentially eroded away over time and a faint pseudo negative image with some 3-dimensional properties was left where the acid in the original slurry medium chemically reacted with a superficial layer of the linen cloth.

5. Finally, the major bloodstains were painted on the cloth and a pen-sized butane blowtorch was used to mimic the large burn scars from the 1532 fire.

F7	**Proto-Photograph Hypothesis (Allen)** [1][2][3][4][5][6]

In the mid-1990s an ingenious hypothesis for formation of the image was proposed and tested by South African researcher Nicolas P. Allen. The hypothesis was in part motivated by Allen's study of all previously proposed artistic mechanisms and their failure to explain the image. Allen recognized that the image in many respects had the characteristics of a photographic negative and concluded that the best explanation for the image was, in fact, an early photography-like method from the Middle Ages. In the published paper on his work Allen stated:

> "...*if one considers the facts as they are presented by the Shroud as Sache selbst* (German: the thing itself), *it would seem that the only possible and logical way that the image on the Shroud could have been produced was by a photographically-related technique.*" [7]

Allen's method of producing an image was to employ a camera-obscura method. He used a life-size sculpture of a human body lit by sunlight and focused the image of the illuminated sculpture through a lens made of quartz, approximately six inches (15.3 cm) in diameter, into a dark chamber that contained a chemically treated and thus photographically-sensitized cloth. Allen executed the process in the following steps:

1. A linen cloth was treated with silver nitrate to make the cloth sensitive like a photographic film.
2. The frontal image was captured first.
3. The set-up was changed to capture the dorsal image second.
4. The set-up was changed again to separately capture a facial image.
5. The cloth was washed to remove any residual silver nitrate.

Allen was able to produce a compelling image on the cloth that showed a photographic method could create an image with many Shroud characteristics. However, when his results were studied carefully many questions were raised. Ultimately, the hypothesis was unable to gain much traction mainly because of one major problem pointed out by photographer Barrie M. Schwortz, the well-known STURP photographer. Schwortz makes the point that the historical context of Allen's hypothesis is flawed because no medieval examples of his technique have ever been found. [8] Allen was not able to produce one example of his method from any medieval (1260-1355) source. If the Shroud image is a proto-photograph, it is the only one known to exist from that era. If the Shroud existed prior to 1200, in accordance with the weight of historical evidence, then the probability of Allen's method being used would not only be unreasonable, it would be vanishingly small. Besides the historical incongruity of Allen's hypothesis, there are several inconsistencies with regard to the full context of the image characteristics.

F8	**Shadow Hypothesis (Wilson)** [1]

Nicolas P. Allen's proto-photograph hypothesis reinforced the idea that light played a potential role in the production of the Shroud image. The use of the sun, in particular, inspired a rather simple hypothesis by Nathan D. Wilson known as the **Shadow Shroud** hypothesis. Wilson says that his method is quite different in its details from Allen's but that there is something in common with it: "*We are both attempting to create a photo negative by means of sunlight.*" [2]

The Shadow Shroud method is simple and ingenious and it can produce an image that is both pseudo 3-dimensional and shows photonegative attributes. Wilson's method starts with a raw piece of linen that is uniformly "aged" to simulate the maximum image intensity and color of the Shroud image. It hasn't been specified how Wilson "aged" the linen he used or if it was just a sample of old linen that had naturally aged. The method entails using a painted piece of glass whereby the painted image on the glass casts a shadow, proportionally to the density of the paint applied to the glass, onto the cloth, and thus, protects the cloth beneath the painted image. The use of the "shadow" in the image is not to create the image but to protect an already colored cloth while the sun does its work to bleach the "unprotected" cloth. The result is an image created in a way analogous to a sculpture. In a sculpture one can say that the image already resides in the rock, and a talented artist reveals it by removing the material that "hides" the image. Analogously, in Wilson's method the image is already in the colored cloth and simply needs to be revealed by removing what the artist does not want. The instrument is not a chisel and a mallet, but sunlight. The method is able to produce an image, and the image can be as good as the artist who does the painting of the "shadow" on the glass. Wilson included only a facial image in his experimentation, and he did not attempt to include any blood details. Wilson's results are interesting, but there remain insurmountable inconsistencies with the actual Shroud images of a dead and tortured human body.

Radiation/Electric Field

F9 Radiation: Fall-Through Hypothesis (Jackson) [1][2]

John Jackson and his TSC research team have proposed a radiation-based image-formation hypothesis that is theoretically consistent with all of the Shroud image characteristics listed in Section 4. The hypothesis is known as the "Radiation Fall-Through Hypothesis". It was first proposed in 1989 and has been worked on and refined ever since. The unique and unusual 3-dimensional characteristic of the Shroud image inspired Jackson to begin his work on an image-formation hypothesis. In their original work of analyzing the "3-dimensional" phenomenon of the Shroud image, Jackson and his colleagues established that a very close correlation could be established between the intensity of the image and the vertical distance to a hypothetical body wrapped in the Shroud. Experiments with human volunteers established that the cloth-to-body distance correlation was in a vertical direction that appears to be related to the earth's gravitational field (see Items B3 and B5). This fact led Jackson to conclude that gravity was a deciding factor in determining several of the Shroud image characteristics. Jackson's team also conducted experiments using ultraviolet light to irradiate samples of linen cloth followed by heating the cloth in an oven to cause artificial aging. It was found that the irradiated and artificially-aged linen samples developed a superficial colored layer that both visually and chemically closely matched the colored image-bearing fibers of the Shroud. The detailed and complex "Radiation Fall-Through Hypothesis" followed. The hypothesis, in brief, states that the body wrapped in the Shroud became volumetrically radiant (radiant throughout its entire volume) with light in the vacuum ultraviolet range (VUV) and simultaneously mechanically transparent, thus offering time-decreasing resistance to the cloth as it collapsed through the body space under the influence of gravity. Finally, the hypothesis proposes that the irradiated cloth, over some indeterminate period of time, aged and the image developed.

This hypothesis posits a singular event that has been modeled theoretically and through computer simulation, but it clearly cannot be physically replicated. Nevertheless, the hypothesis does make predictions concerning image characteristics that can be evaluated and, ultimately, tested by the scientific method. It is important, in particular, to note that in the process of developing the Fall-Through Hypothesis Jackson predicted that there should be traces of a "doubly superficial" image associated only with the frontal image. This prediction was based on the assumption that as the cloth hypothetically collapsed through the radiant body, the image-forming vacuum ultraviolet light would also irradiate the back of the cloth. This prediction was made while the Holland backing cloth still covered the back of the Shroud in its reliquary in Turin. At the time of Jackson's prediction of a "doubly superficial" image, the back of the Shroud had not been viewed for hundreds of years. When the backing cloth was removed and the back of the Shroud was studied during the 2002 preservation project, as predicted by Jackson, a faint superficial image of the face and possibly of the hands was observed on the back of the Shroud (see Item C3 and C4). This demonstrates the powerful analytical and predictive strength of the Fall-Through Hypothesis.

Antonacci Version of the Fall-Through Hypothesis [3]

Shroud scholar Mark Antonacci in his 2015 book entitled "Test the Shroud" both endorsed Jackson's hypothesis and offered an intriguing variation. Instead of a phenomenon involving vacuum ultraviolet light, as Jackson has proposed, Antonacci and a team of collaborating scientists proposed *particle radiation* consisting of protons and neutrons as the source of the body image and other phenomena associated with the Shroud. Thomas J. Phillips of the High-Energy Physics Laboratory at Harvard University first proposed the idea that a neutron flux phenomenon might be associated with the formation of the Shroud body image. Phillips made his proposal at the time the 1988 Shroud carbon dating results were announced because it is known that a neutron flux can skew the C-14/C-12 ratio of a linen cloth and, thus, cause erroneous carbon dating results (see Dating the Shroud discussion on Enhanced Contamination). Antonacci's team has taken the idea of a neutron flux to construct a complete hypothesis that can be tested through specific tests of Shroud chemistry. Following Phillips, Antonacci's team proposed that if particle radiation was the source of the Shroud image, then not only would an image of the body have been left on the cloth, but unstable isotopes should have been formed in the process and several of these isotopes have half-lives long enough that they would still be present on the cloth, yet short enough that they are not found in nature. Specifically, it has been proposed that if the hypothesis is correct, then rare radioisotopes of chlorine (Cl) and calcium (Ca) should be able to be detected on the Shroud; thus, the title of Antonacci's book: **Test the Shroud at the Atomic and Molecular Levels.** Antonacci has named his hypothesis "The Historically Consistent Hypothesis." In his book Antonacci makes the following summarizing statement regarding his team's hypothesis:

"Only hypotheses involving a burial cloth collapsing into a radiant region once occupied by a disappearing (or mechanically transparent) body can account for all the Shroud's primary and secondary body image features, its skeletal features and its outer side imaging. However, only the Historically Consistent Hypothesis, involving a burial cloth collapsing into a field of particle radiation that consists of protons and neutrons emanating from a disappearing (or mechanically transparent) human body, can also account for the Shrouds still-red color of its blood marks; it's possible coin and flower images, its excellent condition; and its aberrant medieval radiocarbon dating." [4]

In Section 6 where the various image-formation hypotheses are rated, Jackson's vacuum ultraviolet light (VUV) version of the Fall-Through Hypothesis is used. Jackson's hypothesis is mature and has continued to gather more and more support through the years, including from Antonacci's scientific team. The next step for the intriguing Antonacci version of the Fall-Through Hypothesis must be to prepare a detailed testing proposal and submit it to the Shroud custodians. If approved, the proposed testing can be done to attempt to confirm the presence of rare radioactive isotopes of Cl and Ca on the Shroud. Confirmation of their presence would be a landmark discovery.

F10	**Electric Field: Corona Discharge (CD) Hypothesis (Fanti)** [1][2][3][4]
	Shroud scientist Giulio Fanti, building on work conducted by Oswald Sheuermann in the early 1980s, has developed a hypothesis that points to a **corona discharge (CD)** phenomenon being responsible for the formation of the Shroud image. A **CD** is an electric discharge appearing on and around the surface of a charged conductor, in this case the Shroud body shape, caused by the ionization of the surrounding air due to the presence of a strong electric field. The primary threshold of **CD** can be lowered in the presence of radon that also can ionize the ambient atmosphere. Radon, atomic number 86, is an invisible, radioactive gas that is often found in confined spaces in the vicinity of Jerusalem, such as a basement or tomb. Radon concentrations may be increased in the ambient surroundings by an earthquake. During a **CD,** there is normally light emission, mainly in the ultraviolet range due to atomic excitation. A **CD** phenomenon can also produce chemical byproducts that can include ozone, nitric acid and other reactive species that might have a role in image-formation. Fanti finds support for the **CD** hypothesis in the fact that an earthquake and its aftershocks can cause an electric field surrounding the compressed rock layers of Jesus' tomb. Also, the Gospel of Matthew (28:2) hints at the occurrence of a strong earthquake while the body of Jesus was entombed. **CD** phenomena have, in fact, been observed and scientifically documented at the time of earthquakes and in the presence of high concentrations of radon. Fanti's research team has conducted extensive experiments to test the ability of a **CD** phenomenon to produce a Shroud-like image on cloth. Fanti's team used a half-scale metalized mannequin covered with a cloth in a high-energy electric field that produced a **CD** phenomenon. The experiments of Fanti's team are the most extensive conducted by any research team to date. They were able to produce a superficial body image with some 3D characteristics and with double superficiality effects on the cloth although the image itself was very distorted.

Note: In 2012, a team of scientists at the ENEA Research Centre in Frascati, Italy led by P. Di Lazzaro, in collaboration with Dr. D. Murra and Dr. A Santoni, published an important paper in **Applied Optics** entitled "*Superficial and Shroud-like coloration of linen by short laser pulses in the vacuum ultraviolet*". This paper offers important findings that may have relevance in relation to Jackson's Radiation (vacuum ultraviolet light) hypothesis and Fanti's **CD** hypothesis. [5]

Electric Field: Electric Charge Separation (D.S. Spicer and E.T. Toton) [6]

This hypothesis, like Fanti's, relies on the presence of an enveloping electric field in a confined tomb during an earthquake. The hypothesis suggests that sweat on the body may be the vehicle for the formation of the Shroud image. Sweat contains urea, which is acidic and can cause cellulose to yellow. Also, the urea molecule has a large polar moment that might cause it to attach to the molecular charge exposed on the outer surface of a very poor conductor of electric current (dielectric), such as the Shroud cloth. Spicer and Toton emphasize a **low-energy electric field** in their hypothesis. They rely on reactive polar molecules diffusing from the body, which are then concentrated by electric fields at locations on the body, thus causing the body geometry to be mapped to the surface of the enveloping cloth. This hypothesis, though very interesting, has not yet been experimentally tested, which currently limits detailed evaluation.

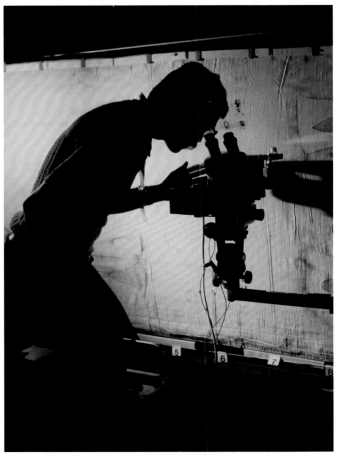

(Fig. 92) Mark Evans examines the Shroud with a high-powered photographic microscope during the 1978 STURP scientific expedition. The STURP findings only acted to deepen the mystery of the image-formation process. The image remains *Acheiropoieta*, the term first applied to the *Image of Camuliana* in ca. 540 AD.

SECTION 6

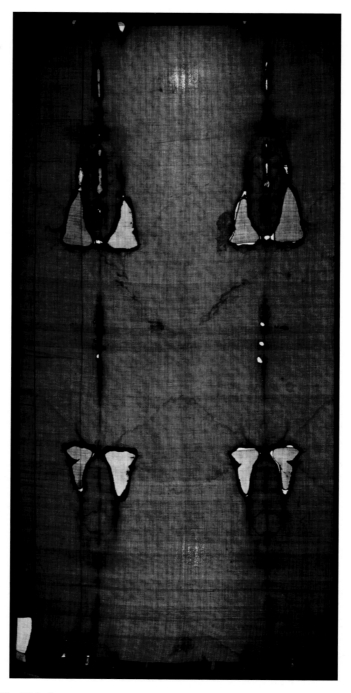

(Fig. 93) This is a transmitted-light photograph of the Shroud; that is, the illumination is from behind the cloth. Notice how the body image seems to disappear with only the blood stains being prominently visible. Also note that this is a direct photograph of the Shroud with the bloodstain from the chest wound on the right side of the image.

Academy of Sciences or the Archbishop of Turin, should oversee the testing effort; how many testing laboratories should be involved; how many samples should be taken?

Finally, a meeting was convened (29 September–1 October 1986) in Turin to establish a protocol for the radiocarbon dating of the Shroud. Carlos Chagas, the head of the Vatican's Pontifical Academy of Sciences at that time, chaired the meeting. He was operating on the assumption that the Vatican Academy of Sciences would participate throughout the radiocarbon dating process. [7] The Turin authorities, on the other hand, believed they should control the process, and they insisted that certain specialists that they designated be invited to attend the meeting. Among others, Turin nominated the chemist Alan Adler, who had been a member of STURP, and the Hong Kong archaeologist William Meacham. Both were subsequently invited to the meeting in spite of efforts by the radiocarbon contingent to have them excluded. [8] The protocol that was hammered out at the meeting came to be known as the **Turin Protocol**, and it included the following important agreements: [9]

- For statistical purposes, it was decided that seven carbon dating laboratories would be involved. The laboratories would represent the newer AMS testing method (five laboratories) and the older Small-Decay-Counter method (two laboratories).

- Seven (7) samples would be taken from the Shroud, one for each laboratory. It wasn't specified where these samples would be taken, but Adler stated at the meeting that at least three (3) different areas must be included. Meacham supported him, and both believed at the time that this was the consensus agreement that would be included in the protocol that was published and would be implemented in any subsequent testing. It turned out that they were both wrong in this assumption.

The protocol meeting did not resolve the issue of whether the Archbishop of Turin or the Vatican's Pontifical Academy of Sciences would oversee the radiocarbon dating efforts. The Archbishop, based on the fact that the Shroud was in his immediate custody and that his team in Turin had successfully provided oversight of the 1978 STURP scientific expedition, petitioned the Vatican to have Chagas's Pontifical Academy of Science withdrawn from the project. Turin prevailed and the project became theirs alone to oversee. Subsequently, Turin authorities unilaterally acted to modify the Turin Protocol to reduce the number of testing laboratories from seven to only three, all of them being AMS laboratories (Oxford, Zurich, Arizona USA). Thus, the Small-Decay-Counter method laboratories were excluded. All of the seven original laboratories vigorously protested the making of such arbitrary changes to the Turin Protocol. In a letter sent to the Archbishop of Turin they stated:

". . . we would be irresponsible if we were not to advise you that this fundamental modification in the proposed procedures may lead to failure." [10]

Harry Gove, the key leader of the radiocarbon laboratory contingent, drafted a second letter to be sent directly to the pope. The draft included the following statement:

"Rather than following an ill advised procedure that will not generate a reliable date but will rather give rise to world controversy, we suggest that it would be better not to date the Shroud at all." [11]

The letter to the Pope was never sent. The three laboratories designated by Turin to be involved (Oxford, Zurich, Arizona) would not sign the letter.

The Shroud Sample

Finally, under the direction of the Turin authorities the day came, April 21, 1988, to cut the samples from the Shroud for the three laboratories chosen to do the actual testing.

- Two "qualified textile experts" who were invited by Turin to help with the taking of samples apparently had little or no expertise concerning the Shroud, and were reportedly seeing the Shroud for the very first time. [12]

- After what was reported to be two hours of debate to finally decide on the location for the sampling, only ***one*** sample was cut from the Shroud. [13] The spot selected was near the hem where the Raes sample had been removed in 1973, exactly the location that Meacham had issued warnings about two years earlier. The single sample cut from the Shroud was subdivided into pieces and distributed to representatives of the testing laboratories. There is no record of the senior representatives of the three testing laboratories (Oxford, Zurich, Arizona), who were present in Turin and observed the taking of the sample, making any objections to the procedures used to select the location of the sample or that only one sample location was chosen.

- The three laboratories apparently used their extensive but "standard" cleaning methods for the pieces of the Shroud they received for testing. [14] They did not report doing the *"elaborate pretreatment, scanning electron microscopy (SEM) screening and testing (micro chemical, mass spectrometry, micro-Raman) for impurities or intrusive substances such as higher order hydrocarbons, inorganic and organic carbonates"* recommended by Meacham.

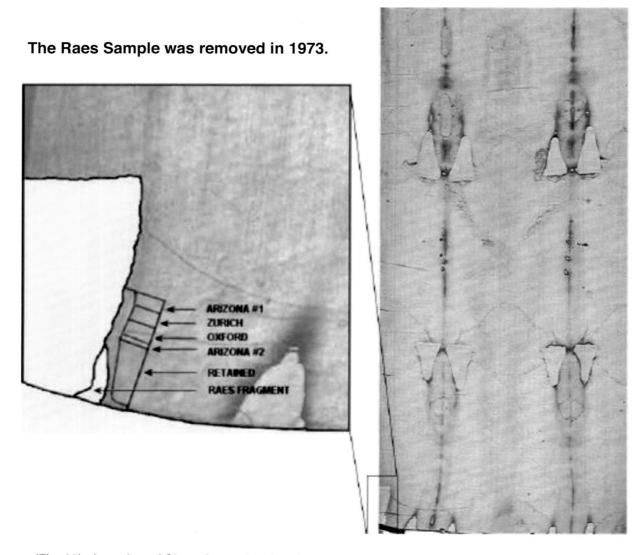

(Fig. 95)   Location of Shroud sample taken for radiocarbon testing relative to Raes sample area

Radiocarbon Dating Results Announced

On October 13, 1988, at a press conference that the world anxiously awaited, Cardinal Ballestrero of Turin announced the official results: The radiocarbon dating tests returned an age for the Shroud of 1260-1390 AD, with 95% statistical confidence. The Oxford radiocarbon laboratory presented nearly identical results on the same day, in virtually a simultaneous press conference, in the United Kingdom. [15] The full test results were subsequently published in the 16 February 1989 issue of the science journal *Nature*. [16]

In the whole episode, ***every one*** of Meacham's warnings, along with similar warnings from the STURP team and others, as well as the core provisions of the Turin Protocol, were ignored. The head of the Oxford radiocarbon laboratory, the late Edward (Teddy) Hall, is shown in a photograph taken at the Oxford press conference, in front of a blackboard on which the radiocarbon dating date range for the Shroud is written with a following emphatic exclamation point. He has his arms crossed and a smug expression. Unfortunately, at the press conference Hall said exactly what Meacham thought "no responsible radiocarbon scientist" would ever say. Hall said:

"There was a multi-million-pound business in making forgeries during the fourteenth century. Someone just got a bit of linen, faked it up and flogged it!" [17]

(Fig. 96) Teddy Hall of the Oxford radiocarbon lab (left) , Michael Tite of the British Museum (center), and Dr. R. E. M. Hedges also from the Oxford radiocarbon laboratory (right).

Aftermath

In 2010, the head of the Arizona radiocarbon laboratory co-authored a peer-reviewed paper that gave a description of a retained piece of their Shroud sample that was reported to be still in the laboratory's possession. [18] In the paper the number given for the fiber count of the typical warp thread was 30 and for the average weft thread 40. This is the reverse of that reported by other researchers who have carefully studied the Shroud. [19] Arizona further reported that the thickness of the Arizona sample was 250 micrometers (microns). John Jackson reported that during the STURP expedition that his measurements in micrometers, as recorded by STURP chemist Ray Rogers, were as follows: [20]

Front part of Shroud: 350, 342, 355.

Rear part of Shroud: 391, 358, 348, 362.

Dorsal foot area: 318, 315, 331.

Note that no measurement reported by Jackson and Rogers was less than 300 micrometers (also see Item L2). The question must be asked: Were Arizona's reported measurements simply a reporting error that can be explained as a numerical "round-off" for the fiber count or due to a different technique, or perhaps, as a pressure variation applied to the measuring calipers used for determining the sample thickness? Or was the Arizona sample clearly that different from the main body of the Shroud or the Shroud itself?

In a paper released in 1998, and in a second peer-reviewed paper published in 2013, statistical discrepancies found in the reported results of the 1988 radiocarbon testing were analyzed. [21] [22] These discrepancies, it was shown, could be correlated to the linear location of the pieces that were distributed for testing from the single sample cut from the Shroud. They concluded that due to the heterogeneity of the data and the evidence of a possible linear trend of dates within the radiocarbon sample itself, the radiocarbon tests undertaken in order to date the Shroud could not be considered as a repeated measurement of a single unknown.

In 2015, an alternative analytic method for estimating the age of historic linen fabrics was proposed. [23] A research team in Italy used chemical and fiber tensile-strength comparisons. Specifically, they compared the chemical signature and tensile strength of Shroud fibers they allegedly had in their possession against tables derived from testing a series of "control" linen cloths with known ages, from modern to ancient. Their experiments returned a date of 372 AD ±400 years. The Italian team's results are not yet broadly endorsed. An area of concern is the inherent difficulty of knowing the environmental conditions in which any historic cloth has been kept and the sensitivity of the chemical and structural characteristics of the cloth to these unknown conditions. But the importance of research that searches for an alternative dating method for historic fabrics must be noted and encouragement given for further research in this complex area that is still in its early stages.

TSC Comments

The 1988 radiocarbon testing of the Shroud turned out to be tragic. It was tragic not only because the testing was a flawed scientific exercise, as outlined above, but also because radiocarbon testing was carved out of the 1984 STURP integrated testing proposal. Furthermore, the radiocarbon testing was done **first** ahead of any of the other proposed STURP studies, which meant the data from the proposed STURP studies was not available to critically review the radiocarbon testing result. Then, when the radiocarbon testing results were announced, everything changed. The custodians of the Shroud seemed to become uncertain as to the nature of what they really had in their possession. As a result, they halted all other scientific examinations of the Shroud. None of the STURP 1984 proposed studies, other than radiocarbon testing, were ever conducted, and the Shroud has been virtually inaccessible for scientific analysis ever since. In addition, the Shroud custodians seemingly made the judgment that after the apparent medieval dating of the Shroud, their highest priority became the preservation of the Shroud cloth itself, leading to the 2002 restoration project (see Item H26).

In spite of all the difficulties, from a scientific perspective, TSC accepts the validity of the reported radiocarbon measurement of the C-14 to C-12/C-13 ratio in the tested sample. [24] We presume the three laboratories (Oxford, Zurich, Arizona) conducted their measurements with integrity and correctly measured the relative amount of C-14 in the pieces of the Shroud they tested. We do not accept, however, that the reported radiocarbon date represents the true **calendar date age** of the Shroud. There is a complex of historic, archeological, cultural, and scientific findings that point to a much earlier calendar date for the Shroud. TSC's own empirical study of the fold lines on the Shroud point to a Byzantine history centuries earlier than the radiocarbon date of the fourteenth century (see Item L6). [25] It remains important, however, even with an anomalous radiocarbon testing result, not to simply reject that result but to continue to work diligently to show why it is anomalous in a way that is both prudent and compatible with correct scientific methodology.

We offer the following points related to the search for the cause of the anomalous results:

1. **Radiocarbon Dating Assumptions:** Implicit in the radiocarbon dating process are three fundamental assumptions:

    - The sample or samples tested are representative of the whole.

    - No contamination has affected the C-14 content of the sample area except for the natural radioactive decay of C-14.

    - The initial relative amount of C-14 in the sample is knowable.

2. **Representative Sample:** Significant research has been offered to support the conclusion that the sample area that was tested is not representative of the entire Shroud. Mentioned above are several studies criticizing the statistical results reported by the radiocarbon laboratories. Also, chemical evidence has been offered in support of the non-representative nature of the sample used for the radiocarbon testing. STURP chemist Alan Adler stated:

    "There is far more salt in the radiocarbon [sample] fibers than the water stains on the rest of the Shroud. That is because in the [other] water stains, when the water hit and the soluble salts started diffusing out into the cloth they diffused without limit until they stopped diffusing. But if you look at where the sample was taken it is a 'bounded' water stain. So, all the soluble materials diffuse until they hit that edge and then concentrate there ... [and] ... the radiocarbon [sample] fibers have a different chemical composition from the non-image fibers of the body of the cloth. Therefore, you have a right to raise the question: Is this a representative sample? It doesn't matter whether you think you have an answer to that question or not. It is not a representative sample." [26]

3. **Standard Contamination:** There are two viable, competing hypotheses that invoke a standard C-14 contamination process as a cause for the anomalous testing result:

    Reweave Hypothesis: [27] [28] [29] [30] This hypothesis suggests that the corner of the Shroud from which the radiocarbon sample was cut was repaired with younger materials sometime during the Shroud's history in Europe. While some evidence has been offered in support of the reweave hypothesis there are important counter arguments. Among the latter is the fact that the textile experts who were involved in the 2002 Shroud conservation project examined the sample area and reported they could not identify or find explicit evidence of any reweaving. Also, TSC has carefully studied the X-ray and transmitted-light photographs taken by STURP of the sample area and has seen no evidence of reweaving in this type of imagery. TSC thinks the negative evidence for reweaving is difficult to overcome, and the ultimate answer to the anomalous radiocarbon testing result lies elsewhere.

    Bioplastic Contamination: [31] This hypothesis proposes that living microbes (bacteria or fungus) left a bioplastic coating on the Shroud fibers, at least in the sample area if not on the entire Shroud, and this bioplastic residue explains the anomalous dating result. This theory rests on the commonly understood carbon cycle that holds that atmospheric carbon dioxide is the C-14 source of any contaminating microbe. TSC rejects this standard C-14 contamination hypothesis as an explanation for the anomalous results. To skew the radiocarbon date based on a carbon dioxide source of C-14, from the first century to the fourteenth century date proposed by the radiocarbon tests, would essentially require a near doubling of the mass of the sample by the bioplastic contamination. To the contrary, TSC has shown that the radiocarbon sample, in terms of its mass per unit area, is similar to the average areal density of the rest of the Shroud which itself is similar to any other linen cloth woven to the same specifications. [32] In other words, the mass density for a unit of area of the sample tested is what would be expected of a non-contaminated sample.

4. **Enhanced Contamination:** TSC thinks the explanation for the anomalous radiocarbon date lies with a non-standard or "enhanced" C-14 cause. Two viable hypotheses have also, so far, been advanced for an "enhanced contamination" cause for the anomalous radiocarbon date for the tested Shroud sample.

Carbon Monoxide: The C-14 isotope is formed in the upper atmosphere when cosmic rays generate secondary neutrons that convert atmospheric Nitrogen-14 to C-14 with the expulsion of a proton. In recent years, it has been confirmed through chemical kinetic studies that the C-14 formed in the upper atmosphere does not, as often assumed, initially bond with two oxygen atoms to form carbon dioxide (CO_2). Instead, a predominant fraction of the C-14 bonds with a single oxygen atom in the atmosphere to form carbon monoxide (CO). The carbon monoxide produced typically takes 30-60 days to convert fully to carbon dioxide. As a result, carbon monoxide found at the earth's surface is **highly enriched** in radioactive C-14. TSC's John Jackson has published a hypothesis that this carbon monoxide at the earth's surface might be a significant source of contamination. [33] He pointed out in his paper that given the degree of natural radiocarbon enrichment that has been measured in atmospheric carbon monoxide at sea level, only a small amount of "enhanced" contamination of about 2% carbon relative to the overall carbon in the sample would be required to move a first century date of the Shroud textile to the fourteenth century.

The TSC team has conducted experimental work exploring possible pathways for contamination. In 2008, Jackson discussed preliminary experimental results with the Oxford radiocarbon laboratory. After the discussion, the then head of the laboratory, Christopher Ramsey, made a forceful statement that acknowledged the complexity of the contamination issue. Significantly, this statement erased the emphatic 1988 exclamation point on the blackboard and erased Teddy Hall's rash comments. Ramsey said the following, which is a great credit to him and to the Oxford laboratory, although his statement is not widely known to the general public:

"With the radiocarbon measurements and with all of the other evidence which we have about the Shroud, there does seem to be a conflict in the interpretation of the different evidence. And for that reason, I think that everyone who has worked in this area, the radiocarbon scientists and all of the other experts, need to have a critical look at the evidence that they have come up with in order for us to try to work out some kind of a coherent story that fits and tells us the truth of the history of this intriguing cloth." [34]

Neutron Flux: In the same issue of *Nature* that reported the 1988 radiocarbon testing results, there was an important letter to the editor. This letter rings out today with possibly more force than when it was first written. It causes one again to ponder and adopt a position of caution. The correspondence was with Thomas J. Phillips of the *High-Energy Physics Laboratory* at Harvard University. [35,36,37] Phillips suggested that the Shroud might be a "fundamentally-altered" fabric with respect to its C-14 content due its possible witness to some unexplained event, possibly in the tomb of Jesus. He hypothesized that such an unexplained event, which itself cannot be the subject of scientific inquiry, may have had an effect on the Shroud that can be studied scientifically. The unknown event may have generated a flux of neutrons that could have skewed the C-14/C-12 ratio of the linen cloth. Phillips said that if this were the case, other unstable isotopes should have been formed, and that several of these have half-lives long enough that they would still be present, yet short enough that they are not found in nature. Consequently, searching for these unstable isotopes on the Shroud might be a fruitful line of research, which we understand others are pursuing (Antonacci team. See Item F9).

TSC Conclusion on Carbon Dating

In conclusion, we are left with what appears to be an anomalous radiocarbon date for the Shroud. TSC, for scientific and historical reasons, does not accept that the radiocarbon date represents even an approximate indicator of the actual calendar age of the Shroud.

Today, however, we still do not know the specific cause for the anomalous radiocarbon testing result. It remains important to continue to conduct scientific studies focused on determining why the 1988 radiocarbon testing delivered what appear to be anomalous results for the tested sample. Until and unless the cause for the anomalous radiocarbon date is understood scientifically, TSC strongly recommends that the scientific community and the public not push for another round of radiocarbon testing.

(Fig. 97)   In 2008 at the Oxford Radiocarbon Laboratory, John Jackson (left) discusses enriched carbon monoxide as a possible mechanism of C-14 contamination of the Shroud with Christopher Ramsey, the head of the Laboratory.

Concluding Comments

The **Critical Summary** presents an overview of the key evidence related to the Shroud of Turin, along with comments drawing on TSC's tens of thousands of hours of Shroud research. What should be apparent from even briefly studying the **Critical Summary** is that there is a large corpus of scientific, forensic and historical evidence related to the Shroud. Furthermore, the evidence is interwoven and sometimes difficult to properly interpret. Thus, it is necessary to evaluate Shroud evidence holistically. Some investment of time and effort must be given to grappling with the totality of evidence before any judgment or intellectual commitment to a position of "***authentic or not***" can be made with sufficient rational weight. Dr. Jackson and his TSC associates, after years of intense research following the completion of the STURP project and coupled with the research findings of an ever-expanding body of Shroud scholars, have come to hold the position that the Shroud of Turin is in fact the burial Shroud of Jesus of Nazareth. Others may judge differently or even suspend judgment. That must be respected, so long as it is recognized that a supportable, intellectual position can be reached only after an honest assessment of the totality of evidence.

Regarding specific evidence, there are two important points that merit restating here. First, the conclusion that the Shroud was in Constantinople in 1204 is strongly supported, both historically and empirically. Such a conclusion means the radiocarbon dating of the sample cut from the Shroud to be 1260-1390 AD failed to date the Shroud correctly. Furthermore, such a conclusion acts to strengthen the power of other empirical, historical and iconic evidence that reaches back from 1204 to the first century and the time of Jesus of Nazareth. However, why the radiocarbon dating of the 1988 sample cut from the Shroud did not properly date the Shroud's true age remains a valid, unresolved question that merits continued research. The second point relates to modern image-formation hypotheses that began to be proposed after the release to the public and the greater scientific community of Secundo Pia's revolutionary negative photographs of the Shroud. All artistic means of creating the Shroud image that have been proposed over the past one hundred plus years and those that propose contact, or gases associated with a dead body, must be ruled out because of **multiple inconsistencies** with known image characteristics. Today the class of hypotheses that appears to best fit the image characteristic evidence invokes the action of photon radiation (light) or some other type of radiation. One might say the Shroud image remains an "***impossible***" image. All the evidence indicates the image is truly ***Acheiropoieta*** **(not made by human hands)**.

The "Fall-Through" image-formation hypothesis belongs to the class of hypotheses that invokes the action of radiation or light. Some in our modern era contend that the philosophy of methodological naturalism, which has generally served science well, makes the "Fall-Through" hypothesis itself "impossible". The philosophy of methodological naturalism that guides scientific research holds that reason is limited to acquiring epistemic certainty only on the basis of naturalism. Thus, scientific naturalism chooses not to consider supernatural causes even as a remote possibility for any phenomena or reality. Nevertheless, the Shroud that is arguably the most unique object in existence must be allowed to speak for itself. The "Fall-Through" hypothesis is strictly data driven and is not intended to offer a scientific "proof" of the Resurrection, but only for the Shroud image itself. To the contrary, the Resurrection can never be scientifically "proven." This is fundamentally true because the philosophy of science includes the stipulation to work to "disprove" rather than to "prove". Science rests on hypotheses, many of the most sublime of which, particularly in physics, can never be said to be proven but can only be made stronger through a continuing accumulation of empirical evidence.

And the Shroud? Very unique claims have obviously been made for centuries about the Shroud. In the face of these claims no hypothesis can be dismissed if it offers the best working "fit" to the evidence of the image characteristics. As physicist John Jackson has stated, for the purpose of explaining the Shroud image based on the best "fit", there can be "***no reason to disqualify radiation, specifically vacuum ultraviolet, as a possible mechanism of image-formation***". It must also be noted that the best and latest proposed ***naturalistic*** hypotheses offered to explain the mechanism of image-formation, such as a corona discharge phenomena, are all extreme "forcing" hypotheses that also severely stress the boundary of what can be considered "natural". They, too, are a challenge to the "believable" and border on the "impossible". ***Acheiropoieta.***

As for the "authenticity" position, the ultimate weight of Shroud research can only be gathered to support the position that the Shroud is a true ***instrumental sign***, a sign that providentially points to or reveals some other deeper truth. In this regard, it is noteworthy that in October of 2014 **Version 1.0** of the **Critical Summary** was the subject of a presentation made at the **St. Louis International Shroud Conference**. Before starting the presentation, the presenter asked the assembled group of more than 160 renowned Shroud experts, scientists, and scholars from around the world to raise their hands if they had come to the reasoned judgment based on a dedicated study of the accumulated evidence that the Shroud is the same cloth that wrapped the body of Jesus of Nazareth in the tomb. About two thirds of those in attendance raised their hand. The question was followed by a second question: Had the judgment of "authentic" changed their lives? It was observed that roughly the same hands were raised. There were no follow-ups and no further elaborating discussion. The hands were simply raised.

The making of an intellectual judgment is not mandatory. Even powerful empirical evidence coupled with intriguing historical data that is consistent with a first century Shroud origin do not compel a judgment that the Shroud is the authentic burial cloth of Jesus of Nazareth. **Only** first principles and mathematical proofs compel agreement. All other judgments are "*free*" judgments. Making a judgment of "authentic" for the Shroud based on the current corpus of evidence is clearly a fully justified rational judgment, but it is always a free judgment. A judgment of "authentic", however, inevitably leads to another question: Just who is this man that is enshrouded and for what purpose the trauma, suffering, and sacrifice? Rumination on that question may result in hearing a gentle invitation that may indeed change one's life, as testified by those who raised their hands for a second time in St. Louis.

Thus, we close this document with the same statement with which it was opened:

If the truth were a mere mathematical formula, in some sense it would impose itself by its own power. But if Truth is Love, it calls for faith, for the 'yes' of our hearts."

(Fig. 98) The painting to the left is known as "*THE LIGHT OF THE WORLD*" and dates to 1851-1853, forty five years before Secundo Pia's shocking photographs of the Shroud. The artist of the painting was the Pre-Raphaelite artist, William Holman Hunt. The painting shows a figure of Jesus preparing to knock on an overgrown and seemingly long unopened door. The artist made a point about his painting that expressed his own experience: "*The door in the painting has no handle, and can therefore be opened only from the inside.*" [1] For those who come to the judgment that the Shroud is the "authentic" burial cloth of Jesus of Nazareth they may see it as a sign that further testifies to a faith already held. Others, like Hunt, may simply discern a soft "knock" that evokes the words of Revelation 3:20:

Behold, I stand at the door and knock. If anyone hears my voice and opens the door, then I will enter his house and dine with him, and he with me.

Revelation 3:20

For still others who, after much study and serious contemplation, arrive at the judgment that the Shroud is "*authentic*" a different scripture passage may come to resonate:

Let the hearts that seek the Lord rejoice; turn to the Lord and his strength; constantly seek his face.

Psalm 105: 3-4

Appendix 1: STURP Team Members

Name	Organization	STURP Responsibility
1978 Turin Expedition Team		
John P. Jackson	U.S. Air Force Academy Assoc. Prof. of Physics	**STURP** President/measurements/analysis
Eric J. Jumper	U.S. Air Force Academy Assoc. Prof. of Aeronautics	**STURP** Vice President/measurements/analysis
Joseph S. Accetta	Lockheed Corporation	Infrared spectroscopy
Steven Baumgart	U.S. Air Force Weapons Laboratory	Infrared spectral measurements
Ernest H. Brooks II	Brooks Institute of Photography	Scientific photography
Donald Devan	Oceanographic Services, Inc.	Scientific photography/image analysis
Rudolph J. Dichtl	University of Colorado	Technical support of all experiments
Robert Dinegar	Los Alamos National Laboratory	Chemistry, tape sample removal/analysis
Thomas F. D'Muhala	Nuclear Technological Corporation	Logistics
Mark Evans	Brooks Institute of Photography	Microphotography
John D. German	U.S. Air Force Weapons Laboratory	Technical support for all experiments
Roger Gilbert	Oriel Corporation	Visible/UV spectroscopy
Marty Gilbert	Oriel Corporation	Visible/UV spectroscopy
Thomas Haverty	Rocky Mountain Thermograph	Thermography
Donald Janney	Los Alamos National Laboratory	Image analysis
Joan Janney Rogers	Los Alamos National Laboratory	Technical support

J. Ronald London	Los Alamos National Laboratory	X-ray radiography and X-ray fluorescence
Jean Lorre	Caltech Jet Propulsion Laboratory	Image analysis
Donald J. Lynn	Caltech Jet Propulsion Laboratory	Image analysis Of Note: Lynn was Director of imaging on the Voyager, Viking, Mariner and Galileo projects.
Vernon D. Miller	Brooks Institute of Photography	Scientific photography
Roger A. Morris	Los Alamos National Laboratory	X-ray fluorescence
Robert W. Mottern	Sandia National Laboratory	Image analysis, X-ray radiography
Samuel Pellicori	Santa Barbara Research Center	Visible/UV spectroscopy
Ray Rogers	Los Alamos National Laboratory	Chemistry/tape sample removal/analysis
Barrie M. Schwortz	Barrie Schwortz Studios	Documentation photography
Kenneth E. Stevenson	IBM	Public relations
STURP Members not on Turin Expedition, but who later worked with Shroud Samples		
Al Adler	Western Connecticut State Univ.	Biochemistry/tape sample analysis
Robert Bucklin	Harris County, Texas, Medical Examiner's Office	Medical forensics and analysis
Jim Drusik	Los Angeles County Museum	Conservation
Joseph Gambescia	St. Agnes Medical Center	Medical forensics and analysis
John Heller	New England Institute	Biophysics
Larry Schwalbe	Los Alamos National Laboratory	Physics/X-ray fluorescence
Diane Soran	Los Alamos National Laboratory	Chemistry/Archaeology

Appendix 2: Rating Details for Image-Formation Hypotheses

F1 Contact Hypothesis (Vignon)

C1 Frontal/Dorsal Same Max. Density:
Inconsistent: The weight of the body and gravity affecting any decomposition liquids or aromatic embalming oils would inevitably leave traces of discernable differences in image intensity between the frontal and dorsal image. Such differences are not observed on the Shroud.

C2 Superficial Image:
Inconsistent: The image on the Shroud is remarkably uniform in its superficiality over its entire extent. It is judged to be inconsistent to achieve this uniform superficiality with a wet body, whether the body is artificially coated with some type of aromatic embalming oils or if the body is itself producing decomposition liquids. In either case the liquids or oils would penetrate into the cloth through capillary action and thus cause colored fibers below a superficial surface layer.

C3 Superficial Image Backside of Frontal:
Inconsistent: A contact mechanism is incompatible with an image on the opposite side of the cloth from the body image unless the area between the two surfaces is also colored with materials soaking through the full thickness of the cloth. To the contrary, on the Shroud the body image is superficial on both surfaces with no coloring of the middle of the cloth between the front and back surface.

C4 No Superficial Image Backside of Dorsal:
Consistent: A direct contact image-formation mechanism is consistent with this image characteristic. In this case, what is not observed does not need an explanation.

C5 Image Fiber Chemistry/No Paint:
Questionable: It has not been demonstrated that chemicals associated with a dead human body can cause molecular changes to linen fibers similar to that observed on the Shroud.

C6 No Cementation/No Capillary Flow:
Inconsistent: Vignon hypothesized a dead body where decomposition liquids, and aloes on the body were involved in the image-formation process. It is judged to be inconsistent that such a contact mechanism involving materials on the body sufficient to produce an image would fail to leave any microscopic evidence of cementation or capillary flow.

B1 Reverse/Negative:
Consistent: A direct contact image-formation mechanism is judged to be consistent in principle with this image characteristic.

B2 High Resolution:
Consistent: A direct contact method of image-formation can possibly, under ideal conditions, result in a relatively high-resolution image.

B3 3-Dimensional:
Inconsistent: With respect to the frontal body image a natural contact mechanism cannot account for the closer to body=denser nature of the image on the Shroud. With a natural un-manipulated contact method, you would tend to get an all or nothing binary imaging effect. If the body was in contact with the cloth you would have an image. If the body was not in contact with the cloth, then no image would be imprinted. Jackson, Jumper and Ercoline performed experiments with a direct contact mechanism and concluded, "*such binary characteristics pose a fundamental problem with this type of process (natural body contact), for such behavior does not provide the necessary latitude to correlate intensity with a continuum of relief variations.*"[1] It is also important to note that STURP scientists found that the Shroud image is continuously shaded, at least to some degree. There are no regions over the entire body image area where absolutely no colored image fibers are to be observed except where there are bloodstains or wound exudates.

B4 Wrapping Distortions:
Inconsistent: Consider the Shroud draped over the underlying body. There are three (3) primary ways for the points on the body to be mapped to the cloth that is consistent with the requirement of producing a high-resolution image:

1. Mapping perpendicular to cloth.
2. Mapping perpendicular to the surface of the body.
3. Mapping in a vertical direction congruent with the direction of gravity.

When the Shroud with its image is laid flat, generally corresponding to what is seen in photographs, there inevitably will be 2-dimensional image distortions that can be correlated with the mapping phenomena. It has been demonstrated that the 2-dimensional image distortions on the Shroud are consistent with a vertical, in line with the direction of gravity, mapping of body points to the cloth. A contact mechanism maps only contact points between the cloth and the body. The 2-dimensional distortions from a contact image-formation mechanism are different from those obtained from a vertical mapping phenomena and can be critically distinguished from them.

B5 No Body Side Images:
Inconsistent: There is no experimental evidence to demonstrate that in a direct contact method it is possible to prop up the sides of the Shroud away from the body with foreign objects such as spice bundles or flowers in such a way as to achieve the precision needed to preserve the anatomical correctness of the body and the absence of side images that is observed on the Shroud. The blood in the hair area is consistent with the Shroud picking up these images by contact with the face while the Shroud was wrapped around the head, not propped up away from the head and sides of the face. Yet there are no side images of the face, only the blood.

B6 Blood and Serum:
Consistent: The human blood, wounds and serum complex found on the Shroud is consistent with the Shroud enfolding a tortured and dead human body

B7 Off-Image Blood:
Consistent: A contact hypothesis would appear to be generally consistent with the off-image blood. However, there may be difficulty explaining the mis-registration of the body image with the blood in the hair. As mentioned in B5, there are images of blood consistent with the Shroud picking up these images by contact with the face while the Shroud was wrapped around the head. But if the blood is transferred from the sides of the face why are there no facial side images? While noting this apparent inconsistency, we still judge the off-image blood to be consistent with a contact mechanism.

B8 No Image Under Blood/No Imaging Damage:
Consistent: The blood complex was deposited onto the Shroud cloth before the action of the image-formation mechanism. A direct contact image-formation mechanism is consistent with this requirement at least in principle, although the superficiality nature of the image also must be taken into account. This item is consistent only if the direct contact method does not produce linen fiber coloring beneath blood and serum through capillary action that in a direct contact method might be hard to avoid.

B9 Dead Human Body:
Consistent: The contact hypothesis is directly based on the presence of a dead human body at the time of image formation.

B10 No Putrefaction:
Questionable: A direct contact method of image-formation with a dead human body would likely involve the presence of decomposition products. No decomposition products have been observed on the Shroud.

B11 Bone Structure:

Inconsistent: An image-formation method involving contact with the outside of a dead body would appear to be inconsistent with the imaging of any internal skeletal structure.

F2 Gas Diffusion Hypothesis (Rogers)

C1 Frontal/Dorsal Same Max. Density:
Consistent: Rogers' hypothesis includes the assumption that the body was removed from the Shroud before decomposition liquids formed. This assumption means it is possible in his hypothesis that there would be no observable difference in the Frontal/Dorsal maximum image density.

C2 Superficial Image:
Consistent: Rogers' hypothesized that amine gases reacted with a superficial starch contamination layer left on a micro-thin evaporation surface on the Shroud. If it is granted that he was correct in his starch contamination theory, then a superficial image could be achieved through the Maillard reaction proposed in his hypothesis.

C3 Superficial Image Backside of Frontal:
Questionable: The frontal double superficiality has been reported to include the hair and only minor body imaging of the facial and possibly hand areas. Questions remain concerning the ability of a heavier-than-air gas to diffuse through a covering cloth and react with surface fibers on the topside of the cloth. An experimental demonstration is required to answer this question.

C4 No Superficial Image Backside of Dorsal:
Inconsistent: Rogers' hypothesis included a conclusion that the amine putrefaction gases, although being heavier than air, would diffuse through the upper portion of the linen cloth covering the body and produce a doubly superficial image on the top of the Shroud of at least the hair. He theorized that the hair around the face would be a natural "trap" for high quantities of gas and that the gas would diffuse from the hair through the top of the cloth and color fibrils on the backside. This explanatory power of this theory, though untested, remains feasible. The dorsal side of the image and the cloth below the body presents a bigger problem for the gas diffusion hypothesis. The amine putrefaction gases from the body should also diffuse through the cloth below the body. In this case, the gases are aided by gravity and are confined beneath the cloth by any object the body is supported by, such as a stone slab. It is very unlikely that there would be convection currents under the body that could carry away the relatively heavy amine gases. Consequently, the backside of the cloth below the body should also be expected to exhibit color from the proposed Maillard reaction. To the contrary, no such coloring on the backside of the cloth beneath the dorsal body image has been found.

C5 Image Fiber Chemistry/No Paint:
Inconsistent: Roger's hypothesis proposed that the image fibers received their color through a Maillard chemical reaction between amine gases, and possibly ammonia gas, and a starch contamination layer on the evaporation surface of linen fibers of the Shroud. Heller and Adler in their paper "*A Chemical Investigation of the Shroud of Turin*" [2] could not identify any starch being present on the Shroud. Rogers reported, however, that spot tests with aqueous iodine indicated the presence of some starch fractions on Shroud fibers. But even if there was a starch layer on the Shroud, water would tend to partially dissolve some of the carbonyl compounds and aromatic substances that are formed in a Maillard reaction and that are water-soluble which in our judgment would leave behind discernable trace evidence. There are significant ancient water stains on the Shroud (see Item L10) and water stains associated with the quenching of the 1532 fire. The image fibers in the water stain areas of the Shroud are not observed to be affected. STURP tests also confirmed the colored image layer on the Shroud linen fibers was not water-soluble.

C6 No Cementation / No Capillary Flow:
Consistent: The gas diffusion mechanism proposed by Rogers is consistent with this image characteristic. In his hypothesis, the body was removed from the cloth before any decomposition liquids formed on the body.

B1 Reverse/Negative:
Consistent: The gas diffusion image-formation mechanism proposed by Rogers is consistent with this image characteristic.

B2 High Resolution:
Inconsistent: The areas between the legs and between the cheeks of the buttocks on the Shroud dorsal image offer a simple example of why a gas diffusion model with a heavier-than-air gas such as Rogers proposes appears to be inconsistent with the Shroud image. The heavier-than-air gas produced by a dynamically gas producing decomposing body would saturate the area between the legs and cheeks of the buttocks beneath the body and cause this area to be densely colored. To the contrary, the separations between the legs and cheeks of the buttocks in the Shroud dorsal image are clearly resolved and are not densely colored. Rogers was careful and clearly stated that "**the important point to recognize is that blanket, qualitative statements about diffusion and resolution can not be supported by simple assumptions.**" [3] However, Rogers published no experimental results for image resolution on a cloth laid out **below** an amine gas generator. We are not aware that anyone else has either. The bottom line is that there appears to be no published evidence that a gas diffusion model based on a decomposing body producing heavier-than-air amines can produce the frontal and dorsal body image resolution observed on the Shroud. All macro evidence available today is to the contrary. Experimental evidence would need to be presented and carefully scrutinized to show that high resolution can be obtained with a heavier than air gas diffusion mechanism, or any gas diffusion mechanism for that matter, particularly beneath the body (see comments below in B3 that also reflect on high resolution in relation to a diffusion image-formation mechanism).

B3 3-Dimensional:
Inconsistent: The three-dimensional nature of the Shroud image, particularly the frontal image, is due to the difference in the density of the image being correlated to cloth-body distance. A diffusing gas model would seem to have no means of conveying this information to the cloth. Experiments conducted to date to model a gas diffusion image-formation mechanism all show grossly distorted results. Among other factors that present major hurdles is the factor of time. Rogers has demonstrated that amines, heavier than air, should be able to react in a Maillard reaction with a proposed starch impurity layer on a linen cloth to produce a Shroud-like coloring of individual linen fibers. This is not the fundamental problem with the hypothesis, assuming the contamination layer theory is correct. One major problem is the continuous dynamic nature of a gas diffusion model. The body will keep producing amines, potentially increasing over some period of time for the entire body, leading to saturation of the image area in such a way as to homogenize the intensity structure thus degrading resolution and any subtle differences in cloth-body distances. Rogers does not appear to have fully addressed this dynamic of continuing amine production in experiments associated with his hypothesis. Jackson, Jumper and Ercoline experimented using other approaches to a diffusion mechanism. Their initial studies of the diffusion mechanism were performed by soaking a plaster reference face, as shown below, in an ammonium hydroxide solution and then draping a cloth sensitized with mercuric nitrate over it, noting that the attempt was not to simulate image chemistry, but only image structure defined by the physical aspects of the diffusion process. Reaction of ammonia vapor gave a brownish discoloration that constituted the image.

Inconsistent: Garlaschelli used an actual human body for acquiring the frontal and dorsal body images. However, he apparently had the model simply turn over onto his stomach to create the dorsal image. This however wouldn't work in a simple fashion, as the model would have to assume the same relative posture for the frontal and dorsal images (like a true bas-relief). This would be very difficult. The facial image was made using an actual shallow bas-relief and the same comment for hypothesis F5 applies.

B5 No Body Side Images:
Consistent: A frottage method can achieve this result.

B6 Blood and Serum:
Inconsistent: A frottage method is an artistic method. Therefore, the same comment as for the painting hypothesis F3 applies.

B7 Off-Image Blood:
Questionable: A frottage method is an artistic method. Therefore, the same comment as for the painting hypothesis F3 applies.

B8 No Image Under Blood / No Imaging Damage:
Inconsistent: Garlaschelli openly stated that he used paint to mimic blood and wounds as a second step after image formation. Clearly in his method there are colored image fibers under the painted blood. But even if he had attempted to do otherwise the same comment as for item F5 applies. The only difference in Garlaschelli's method from the F5 bas-relief method is that the damage to underlying blood would not be caused by heat but by acid present in his image-making slurry.

B9 Dead Human Body:
Inconsistent: The use of a dead body to create blood and wound details is inconsistent with Garlaschelli's bas-relief method, even if he had attempted to do so. This is because there is no acidic damage to the blood or serum retraction rings observed on the Shroud.

B10 No Putrefaction:
Consistent: A frottage method is an artistic method. Therefore, the same comment as for hypothesis F3 applies.

B11 Bone Structure:
Inconsistent: A frottage method is an artistic method. Therefore, the same comment as for hypothesis F3 applies.

F7 Proto-Photograph (Allen)

C1 Frontal/Dorsal Same Max. Density:
Consistent: A camera obscura method can achieve this result.

C2 Superficial Image:
Questionable: Allen's method involves treating a cloth with silver nitrate to make the cloth sensitive to light like a photographic film. A silver nitrate emulsion, or really any emulsion, applied to a cloth would be pulled into the cloth by capillary action and thus would sensitize more than a micro-thin surface. A cloth is simply not like the solid surface of a photographic film.

C3 Superficial Image Backside of Frontal:
Questionable: Is it possible that an artist would even conceive of this detail? To be done with Allen's camera obscura method would require a statue, at least for the head and possibly the hands, to be constructed in reverse left-right orientation to achieve the proper front-back image registration. There is no historical precedent and it is very doubtful that an artist would even consider such a step. In any case, a superficial image on the back of the cloth has the same superficiality problems discussed in C2 above.

C4 No Superficial Image Backside of Dorsal:
Consistent: In the case of the dorsal image the artist need not attempt to produce such a doubly superficial image. In this case, what is not observed needs no explanation.

C5.0 Image Fiber Chemistry / No Paint:
Questionable: Allen's photo sensitizers were silver salts. There is no chemical or spectroscopic evidence for silver species on the Shroud, nor are there any findings of their expected chemical products on the Shroud.

C6 No Cementation/No Capillary Flow:
Questionable: Treating a cloth with a chemical emulsion would theoretically leave some evidence of cementation and / or capillary flow between thread fibers that could be detected under high power microscopic examination. It is not known if Wilson's camera obscura experimental results were examined microscopically.

B1 Reverse/Negative:
Consistent: A camera obscura method can achieve this result.

B2 High Resolution:
Consistent: A camera obscura method can achieve this result. (Note: The resolution of the method is actually superior to the Shroud image and thus Allen's results, which have the actual realism of a photograph can be easily distinguished from the Shroud image. Consequently, some might judge this item to be inconsistent with the Shroud image.)

B3 3-Dimensional:
Inconsistent: Frontal illumination as employed by Allen's method cannot reproduce the subtle lighter and darker areas of the Shroud that can be correlated to distance between a body and a covering cloth. STURP scientist Alan Adler bluntly stated (see Ref-7.1) that the Allen image is "an **albedo** image and will fail a VP-8 test" (the word albedo is derived from Latin *albedo*, "whiteness," or reflected sunlight). The Shroud is simply not a photograph in the sense Allen has hypothesized.

B4 Wrapping Distortions:
Inconsistent: The camera obscura method cannot simulate the Shroud's wrapping distortions. The method employs a flat cloth. There is no geometrical interaction between the cloth and a body to be translated into wrapping distortion information. There is only the interacting of light with a flat cloth. A bas-relief wrapped, or geometrically interacting with a cloth, can result in some wrapping distortions although they are not judged to be able to match the Shroud wrapping distortions. A proto-photograph **of** a bas-relief or statue cannot achieve the subtleties of the actual wrapping distortions of the Shroud.

B5 No Body Side Images:
Consistent: A camera obscura method can achieve this result.

B6 Blood and Serum:
Questionable: In the camera obscura method, the artist might wrap a dead and tortured body in a cloth as a first step in the process. But even in this artistic method the rating for this item is related to how the presence of a dead tortured body being wrapped in the Shroud as a first step is rated. Because we have judged this artistic method to be questionable with respect to the presence of a dead human body (see below), we must also judge this item to be questionable.

B7 Off-Image Blood:
Questionable: A camera obscura method is an artistic method. Therefore, the same comment as for the painting hypothesis F3 applies.

B8 No Image Under Blood/No Imaging Damage:
Questionable: The camera-obscura method is consistent with there being no image under the blood if a dead body was used to create the bloodstains and wound images on the cloth before the imaging process was executed. There remains a question, however, with regard to the chemical sensitizer that is used. There would need to be a demonstration that there would be no damage to the bloodstains and scourge wounds from the photo sensitizing chemicals. This has not been demonstrated.

There is another question as well. This item generally relates to the coordination between the image and the bloodstains and scourge wounds. Comments have addressed the difficulty of other artistic hypotheses to be consistent with what is observed on the Shroud. There is a similar theoretical difficulty with the camera-obscura hypothesis. Yes, the blood and scourge wounds can theoretically be placed on the cloth by enfolding a dead tortured human body. But there is a question regarding the ability of an artist to focus the body image in coordination with those bloodstains on the cloth, both frontal and dorsal. Allen did not address this ability. Recall that because of cloth draping effects the frontal image on the Shroud and related bloodstains have a different height measurement than the dorsal image and related bloodstains. This means a totally different camera obscura setup would be required for the frontal and dorsal images to be brought into register with the bloodstains and wounds. It might even be required to have a different bas-relief or statue for the frontal and dorsal images to create the totality of the Shroud images.

B9 Dead Human Body:
Questionable: The camera-obscura method could in fact have used a dead body to create the blood and wound images. But the use of a dead body can only be consistent with the method if the body and bloodstains can be shown to be in register and that there has been no damage to the bloodstains or serum retraction rings from the photo-sensitizing chemicals. Neither of these requirements has ever been demonstrated (see Item B7 above).

B10 No Putrefaction:
Consistent: A camera-obscura method can achieve this result.

B11 Bone Structure:
Inconsistent: A camera obscura method is an artistic method. Therefore, the same comment as for the painting hypothesis F3 applies.

F8 Shadow (Wilson)

C1 Frontal/Dorsal Same Max. Density:
Consistent: A shadow method can achieve this result.

C2 Superficial Image:
Inconsistent: Wilson started with an aged colored cloth. His image was achieved by sun bleaching of the cloth in areas beyond the shadow cast by his eye/brain/hand coordinated painted image on glass that was placed on top of the cloth. When he was required to address the superficiality question he suggested that his method would require sun bleaching of the "back" of the cloth in order to remove color from threads and fibers through the depth of the cloth to finally leave a superficial image on the front. He further suggested that this approach would lead to the "desired" superficiality on the image side of the cloth. If an artist even knew the "desired" superficiality it is inconsistent to conceive that Shroud-like superficiality might be achieved with "backside" sunlight bleaching over the full extent of both a frontal and dorsal image.

C3 Superficial Image Backside of Frontal:
Inconsistent: This image characteristic by itself creates another insurmountable hurdle for Wilson's hypothesis. Wilson suggested that superficiality on the front of the cloth could be achieved by sun bleaching of the "back" of the cloth. The presence of any residual superficial image on the back of the cloth is inconsistent with this approach.

C4 No Superficial Image Backside of Dorsal:
Consistent: Wilson's proposed method of backside bleaching is, by definition, consistent with this image characteristic.

C5 Image Fiber Chemistry/No Paint:
Consistent: Using a cloth that was properly baked to create an "aged" cloth in Wilson's method could conceivably leave colored image fibers that closely match the Shroud chemistry.

C6 No Cementation/No Capillary Flow:
Consistent: A shadow method can achieve this result.

B1 Reverse/Negative:
Consistent: A shadow method can achieve this result.

B2 High Resolution:
Consistent: A shadow method can achieve this result.

B3 3-Dimensional:
Inconsistent: There are some pseudo 3-dimensional characteristics in Wilson's **shadow** Shroud. The method however suffers from the same inconsistencies as listed for hypothesis F3.

B4 Wrapping Distortions:
Inconsistent: A shadow method is an eye/brain/hand coordinated artistic method. Therefore, the same comment made for the painting hypothesis F3 applies.

B5 No Body Side Images:
Consistent: A shadow method can achieve this result.

B6 Blood and Serum:
Inconsistent: A shadow method is an eye/brain/hand coordinated artistic method. Therefore, the same comment for the painting hypothesis F3 applies.

B7 Off-Image Blood:
Questionable: A shadow method is an eye/brain/hand coordinated artistic method. Therefore, the same comment for the painting hypothesis F3 applies.

B8 No Image Under Blood / No Imaging Damage:
Inconsistent: Wilson's method starts with an aged cloth whose fibers ultimately "become the image". By definition the image fibers will by definition and necessity lie under the blood and wound images.

B9 Dead Human Body:
Inconsistent: Wilson's artistically crafted image is inconsistent with the use of a dead body. The blood and wound details were not painted onto the Shroud. Yet if using a dead body is attempted to create such images there are other complications as outlined above for other image characteristics:

  A. No image under blood. Wilson has no practical means of bleaching out colored fibers beneath the blood. The cloth starts with colored image fibers.

  B. Complexity of weaving image fibers into bloodstains and scourge wounds artistically.

B10 No Putrefaction:
Consistent: A shadow method can achieve this result.

B11 Bone Structure:
Inconsistent: A shadow method is an eye/brain/hand coordinated artistic method. Therefore, the same comment as for the painting hypothesis F3 applies.

F9 Radiation Fall-Through (Jackson)

C1 Frontal/Dorsal Same Max. Density:
Consistent: The hypothesis predicts that the maximum image density will be the same on the frontal and dorsal images. Image intensity is determined solely by contact time of the cloth with the body region. Thus, assuming the radiation event is operative on a time scale less than the time for the upper part of the Shroud to fall completely through the body region, it follows that the interaction times for cloth points, whether initially in contact with the frontal or dorsal surfaces of the body, are equal. Hence, the doses, or image intensities, at those initial contact points should be equal.

C2 Superficial Image:
Consistent: Once the cloth enters the body region, radiation interacts with each cloth fibril throughout the bulk of the cloth from all directions. However, fibrils on both surfaces of the cloth receive a greater dose than those inside because they are unobstructed by adjacent fibrils. These fibrils would probably be highly absorbing to the radiation because the air, which is less dense by nearly three orders of magnitude than cellulose, is assumed to be highly absorbing to account for image resolution. The net result is an exaggerated dose accumulation of the surface fibrils over those inside the cloth leading to a superficial body image. This argument is diagrammed and discussed in Jackson's paper on the Radiation Fall-Through hypothesis.[12] In addition, side shadowing by adjacent fibrils lying on the surface of the threads could lead to the observed selectivity of fibril coloration. A given surface fibril would brown (after normal aging) to a near asymptotic value depending upon the initial dose received, but the overall dose on a given fibril depends in part upon the degree of shadowing by

neighboring surface fibrils. The greater the average dose in a given region of the cloth, the greater would be the relative number of fibrils that would overcome the effects of adjacent fibril shadowing and eventually brown with age. It might also be that different fibrils have different tolerances to browning for a given radiation dose, and this could also contribute to the observed selectivity of fibril browning. Finally, shadowing by the weave structure itself would prevent discolorations from wrapping around a given thread into the interstitial regions of the weave pattern.

C3 Superficial Image Backside of Frontal:
Consistent: As noted above, the superficial nature of the image is explained by the theory. However, the above reasoning leads to one other prediction concerning the superficiality of the image; the frontal image should reside on *both* sides of the Shroud, whereas the dorsal image should reside on only *one* side. The reason is that when the upper part of the Shroud falls into the body region, radiation from the body impinges upon both sides of the cloth. However, in the case of the dorsal image, radiation impinges from only one side because the cloth there never moves into the body.

C4 No Superficial Image Backside of Dorsal:
Consistent: See discussion for Item C3 immediately above.

C5 Image Fiber Chemistry/No Paint:
Consistent: Electromagnetic radiation that is absorbed strongly in air consists of photons in the ultraviolet or soft-x-ray, region that are sufficiently energetic to photo-chemically modify cellulose.[13] Such photons are also strongly absorbed in cellulose over fibril-like distances. Experiments performed by Jackson and his research team have shown that subsequent aging in an oven of photosensitized (bleached) cloth by shortwave ultraviolet radiation produces a browned pattern like the Shroud body image composed of chemically altered cellulose.

C6 No Cementation / No Capillary Flow:
Consistent: A radiation phenomenon of image formation will not cause cementation between adjacent fibers.

B1 Reverse/Negative:
Consistent: The hypothesis produces a reverse negative image correlating a darker image where the cloth body distance is shortest.

B2 High Resolution:
Consistent: As various points on the Shroud intersect different topographical features on the body surface during the collapse process, radiation dose on the cloth begins to accumulate. Because the radiation is assumed to be strongly absorbed in air, radiation effects on the cloth cannot begin until virtual intersection with the body surface occurs. Thus, a one-to-one mapping between a given point on the body to a unique point on the cloth is achieved for all points on the Shroud, which is equivalent to stating that the image is well resolved.

B3 3-Dimensional:
Consistent: The initial draping configuration of the Shroud over a body establishes the initial cloth-body distances. If, then, the Shroud overlying the body falls into the body region, different points on the cloth will intersect the body surface at different times depending upon how far that point was originally away from the body. Thus, each cloth point will receive a radiation dose in proportion to the time that the point is inside the emitting body region. Since that time is inversely proportional to the initial cloth-body distance, it follows that the radiation dose, and hence image intensity, is likewise inversely proportional to the initial cloth-body distance. Correlation of image intensity with cloth-body distance is consistent with the Shroud VP-8 3-dimensional effect.

B4 Wrapping Distortions:
Consistent: The Radiation Fall-Through hypothesis is based on the presence of a dead body and its relation to the image-formation process. The hypothesis is theoretically consistent with this image characteristic.

B5 No Body Side Images:
Consistent: The hypothesis predicts that as the cloth collapses into the body region, internal stresses within the cloth cause it to bulge away from the sides of the body and at the top of the head. Because the radiation is strongly absorbed in air, very little dose is accumulated in the side and upper head regions of the cloth and, hence, no image is visible there.

B6 Blood and Serum:
Consistent: The Radiation Fall-Through hypothesis postulates the presence of a dead body and its relation to the blood and wound images.

B7 Off-Image Blood:
Consistent: In regions of the cloth which fell vertically downward, body and bloodstains should be in register. However, where the cloth is displaced *laterally* as well as *vertically* during the collapse, notably near the sides of the body, we could expect that the body and blood images should be in mis-register. Such appears to be the case for blood originating from the sides of the face but which have been shifted onto the hair images due, presumably, to a lateral movement of the cloth during the collapse. Another possible body/blood mis-registration is at the dorsal foot where the body and blood imprints seem to be somewhat out of coincidence.

B8 No Image Under Blood/No Imaging Damage:
Consistent: The blood and serum on the Shroud provides interference protection for underlying fibers in the Radiation Fall-Through hypothesis.

B9 Dead Human Body:
Consistent: The hypothesis is based on a dead human body.

B10 No Putrefaction:
Consistent: A Radiation Fall-Through phenomena is consistent with this characteristic.

B11 Bone Structure:
Consistent: The Radiation Fall-Through hypothesis predicts the "***possible imaging of internal body structures.***"[14] If the assumed radiation is homogeneously generated throughout the body region, then image intensity would be determined strictly by the length of time that a given part of the cloth is inside the body region. However, if the radiant emission varied with some physical parameter, such as initial mass density, then internal body structures might be convoluted into the general image picture along with the surface features of the body. However, the fact that the surface details of the body appear to dominate the image indicates that the assumed volumetric emission of radiation would have to have been nearly homogeneous. However, many researchers have noted the elongated fingers of the Man of the Shroud, as can be seen from the documentation for image characteristic B11. In the context of the collapse theory, the hand region might be an example where an internal body structure dominated the image, which normally only recorded body surface topography. In particular, the "elongated fingers" are judged to be actual images of the internal bones of the hand extending into the palm region, which, as the cloth passed through the hand region, recorded a greater dose than the surrounding tissue.

F10 Corona Discharge (CD) Hypothesis (Fanti)

C1 Frontal/Dorsal Same Max. Density:
Consistent: A **CD** discharge method can achieve this result.

C2 Superficial Image:
Consistent: A **CD** method has been shown by Fanti's experimental work to be able to achieve this result.

C3 Superficial Image Backside of Frontal:

Questionable: See extended discussion below under 3-dimensional image characteristic (Item B3).

C4 No Superficial Image Backside of Dorsal:
Questionable: If a frontal double superficial image is produced why not a dorsal double superficial image especially at the sides of the dorsal image where there is sufficient air to support a **CD** phenomenon? Fanti's team used a cloth that covered the frontal side of the

experimental manikin only. We think serious problems emerge with respect to the dorsal image.

C5 Image Fiber Chemistry/No Paint:
Consistent: A **CD** method can achieve this result.

C6 No Cementation/No Capillary Flow:
Consistent: A **CD** method can achieve this result.

B1 Reverse/Negative:
Consistent: A **CD** method can achieve this result.

B2 High Resolution:
Questionable: Fanti's experiment appears to have produced relatively high resolution for the image generated by the hands of his experimental manikin. Serious questions remain however about overall resolution, particularly the face. Fanti correctly stated in the paper detailing the results of his experimentation "*It is well known that the complex phenomenon related to the CD distribution law around a variously postured and corpulent body is not simple to study.*" Fanti's experimental results represent a significant step forward but there still remain important questions regarding the ability of a **CD** phenomenon to produce the **totality** of the Shroud image. Because **CD** is a plasma phenomenon where free positive (ions) and negative (electrons) exist in the air-glow region, depending upon the free ion/electron concentrations, a level of electrical conductivity must occur. In general, electrical currents are generated in response to the driving electric field in a manner that tries to negate that field (for a perfect conductor, in fact, electric fields are repelled by this process from its interior causing the external electric field to be normal at the conductor's surface). Moreover, magnetic fields are generated by the induced electrical currents that can further deflect and deform other similar electrical currents and hence the electrical fields that are responsible for the breakdown and where it occurs.

It is not clear how to determine the significance of such electrodynamic/optical phenomena and to what extent instabilities of the plasma might exist. The point is that the **CD** hypothesis relies upon what appears to be an unstable physics with a multitude of special variables that are not easy to determine, control, or predict. Trying to use such a hypothesis to explain an image whose macroscopic intensity pattern is mathematically well characterized in terms of high resolution and a **global correlation** with cloth-body distance (for the frontal image) raises questions and concerns regarding its promise in explaining the Shroud image. (Also see discussion below in Item B3 for additional comments on resolution).

B3 3-Dimensional:
Questionable: The proposed physics of **CD** is described in the paper by Fanti. [15] This hypothesis proposes that a strong electric field, such as might occur during lightning or during an earthquake generated air breakdown immediately above the body, possibly aided by the convergence of electric fields on the surface of the body. In these relatively intense regions of electric field convergence, electrons are given sufficient kinetic energy by the electric fields to become momentarily separated from the atoms by ionization. When these electrons recombine with the ionized atoms, energy is emitted as photons. Because the ionization energies are in the eV range, the emitted photon energies might contain an ultraviolet component that could interact directly with the linen cloth as a surface absorption, owing to the strong attenuation of the ultraviolet in the cloth material.

In 1984, Jackson's research team conducted an experiment to test the general category of electrostatic imaging which we think applies directly to the **CD** hypothesis as formulated by Fanti. [16] In this experiment, the team first made a model of the space between an approximate ½ scale full 3-D plaster reference face and the enveloping cloth. This model was constructed out of paraffin that had been uniformly mixed with carbon to give it an electrical conductivity of approximately 1 inverse ohm per meter. This low conductivity was sufficient to allow Joule heating by the induced electric currents to warm the paraffin model several degrees C so that the resulting thermal pattern could be observed and photographed in infrared (8-14 micrometers). On the outer (i.e. cloth surface side) of the paraffin model which had the geometry of a draping cloth, an aluminum foil electrode was attached by applying to it a thin mist of spray glue. Into the inner surface of the paraffin model (which had the geometry of the plaster 3-D face), a corresponding 3-D plaster face (that had been nickel-plated to make the facial surface electrically conductive) was inserted so as to make close spatial contact of the face and the paraffin model everywhere between the two surfaces. Additionally, between the conductive face and the inner-surface facial depression, a NaCl electrolyte solution was injected in order to ensure uniform electrical contact. Thus, two electrodes were established on opposite surfaces of the paraffin model of the space between the face and the overlying cloth. Next, a D.C. voltage of 37 V at 1 A was placed across the two electrodes for 60 s so that Joule heating by the electric fields would warm the carbon-containing paraffin to a level where thermal imaging of the cloth-side of the image could be observed by an AGA Thermovision system. It was established by a separate experiment that the 60 second warming time used was short compared to the time for thermal diffusion within and on the paraffin model to blur the surface image. By quickly removing the aluminum foil electrode, the resulting heat image was photographed, the best of which is low resolution as shown below (A), along with its distorted 3-D intensity VP-8 rendering (B).

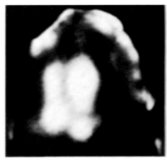 

(Fig. 102)   A   (Fig. 103)   B

In both Jackson's experiment and those conducted by Fanti, (i.e. for a cloth-covered body) the field line structure in the space between the cloth and body obey LaPlace's equation for electrical potentials (see Figure 5 of Fanti's paper), which analyzes the problem via a static electrical model as we did above. At the boundary surface of the body, Fanti assumes that the field lines are normal due to the assumed moistness of the body. This was also the condition of Jackson's experiment because of the high conductivity provided the thin nickel-plate coating over the plaster face. Thus, the field-line structures satisfy both the governing LaPlace's differential equation and the surface boundary conditions required to solve uniquely the electric field structure that corresponds to the specifications of the body-cloth configuration.

It is noted that this case, as argued by Fanti, provides an explanation of the double-superficiality of the Shroud image. However, the resolution of the image by **CD** under this condition where **CD** occurs at the cloth appears to be radically at variance with that observed in the Shroud image itself.

As a point of interest, if **CD** did not extend to the covering cloth but remained in an air breakdown layer over the body surface, then the proposed radiation emitted from this thin layer would emit radiation isotropically, thereby forming an image pattern on the Shroud with specific characteristics. This type of emission was experimentally (and theoretically) addressed by Jackson and colleagues in their 1984 paper with the result that a uniform (featureless) intensity image results.

The essential reason is that each radiating surface element of the body surface emits isotropic R^2 radiation, while each receiving element on the cloth surface sees a surface area on the body that increases as R^2 where R is the cloth-body distance. These two effects cancel leaving

References

Section 1: Historical Evidence

H1
1. Raymond Edward Brown, *An Introduction to the New Testament* (New York, NY: Doubleday, 1997), Table 1, Chronology; xxxviii.

H2
1. Brown, *An Introduction to the New Testament*, 706-7.
2. Flavius Josephus, *The Antiquities of the Jews,* Project Gutenberg Book XIV, Chapter 8, Translated by William Whiston, (January, 2009 EBookl #2848), http://www.gutenberg.org/files/2848/2848-h/2848-h.htm - link142HCH0008.
3. E. Mary Smallwood, *The Jews Under Roman Rule*, (Boston, Leiden: Brill Academic Publisher, Inc., 2001), 37-8, 135, 344-5, 539, 542-43.
4. Jack Markwardt, "Ancient Edessa and the Shroud, History Concealed by the Discipline of the Secret," Proceedings of the Columbus International Shroud Conference, Columbus, Ohio (2008). 6-7. Note: Many theological dissertations were catalogued under the terms *Disciplina Arcani* and *Arcandisciplin*. See *The Discipline of the Secret*, Catholic Encyclopedia, Vol. V, Robert Appleton Company (New York 1907-1914). http://www.ohioshroudconference.com/papers/p02.pdf.
5. Markwardt, "Ancient Edessa." 17.
6. Clement of Alexandria, *The STOMATA*, I,1. http://www.ccel.org/ccel/schaff/anf02.toc.html.
7. Glanville Downey, *A History of Antioch in Syria From Seleucus to the Arab Conquest,* (Princeton, NJ: Princeton University Press, 1961), 273.
8. *The New Interpreter's Study Bible*, (Nashville, Tenn., Abingdon Press, 2003), Commentary for Acts 11:19-26.
9. Tim Dowley, *Eerdmans' Handbook to the History of Christianity,* (Eerdmans Pub. Co., 1977), 62.
10. Downey, *A History of Antioch,* 273-5.
11. Ibid., 275.
12. Maurus Green OSD, "Enshrouded in Silence: In search of the First Millennium of the Holy Shroud," *Ampleforth Journal* 74:3, 321-345, (1969): 329. http://www.monlib.org.uk/papers/aj/aj1969green.htm.
13. St. Jerome, *De Viris Ilustribus (On Illustrious Men),* (The Catholic University of America Press, Washington, D.C., 1999: Translated by Thomas P. Halton), 8.
14. *The New Interpreter's Study Bible*, 2163. The Letter to the Hebrews 9:11 "But when **Christ came as a high priest** of the good things that have come, then through the greater and perfect tent (not made with hands, that is not of this creation).
15. Diana Fulbright, "Did Jesus give his Shroud to "the servant of Peter?", *Proceedings of the International Workshop on the Scientific approach to the Archeiropoietos Images, ENEA Frascati, Italy* (4-6 May 2010). http://www.acheiropoietos.info/proceedings/FulbrightServantWeb.pdf.
16. Jack Markwardt, "Antioch and the Shroud," *Shroud of Turin International Research Conference. Richmond, Virginia* (June 18-20, 1999): 3. http://www.shroud.com/pdfs/markward.pdf.
17. *The New Interpreter's Study Bible*, 2181.
18. Jack Markwardt, personal communication, October 1, 2016.
19. Downey, *A History of Antioch,* Historical Excursus 3, 583-586.
20. *The New interpreter's Study Bible,* Acts 8:14-24, Acts 9:32-35, Acts 10:23-48, Acts 11:1-18, Gal. 2:11-14.
21. John Wehnam, "Did Peter Go to Rome in 42?," Tyndale Bulletin 23 (1972) 94-102.
22. Brown, *An Introduction to the New Testament,* 468.
23. Jack Markwardt, personal communication, September 21, 2016.

H3
1. First Jewish–Roman War, Wikipedia. https://en.wikipedia.org/wiki/First_Jewish–Roman_War.
2. J. Julius Scott, "Did Jerusalem Christians Flee to Pella?," Archaeology Conference, Wheaton College, Wheaton, IL (1998). http://www.preteristarchive.com/Bibliography/1998_scott_flee-pella.html.
3. Jack Markwardt, "Modern Scholarship and the History of the Turin Shroud," *St. Louis International Shroud Conference* (Oct 2014): 57. http://www.shroud.com/pdfs/stlmarkwardtpaper.pdf.
4. Athanasius, *The Sermon of Athanasius,* Mansi, XIII, 584A+Athan.OPP. II353C.

H4
1. Downey, *A History of Antioch,* 272 – 288.

H5
1. Downey, *A History of Antioch*, 215.
2. Smallwood, *The Jews Under Roman Rule, 459-60.*
3. Sister Damian of the Cross (Eugenia Nitkowski, "The Tomb of Christ from Archaeological Sources," *Shroud Spectrum International, Issue 17* (December 1985): 7. http://www.shroud.com/spectrum.htm - 17.
4. Eusebius, *History of the Church, 5, 23, 24.*
5. Markwardt, "Ancient Edessa," 18.

6. Johannes Quasten, *Patrology*, (Westminister, Maryland: Newman Press, Vol. 1, (1950), 171-72.
7. Markwardt, "Ancient Edessa," 1-49.
8. Quasten, *Patrology*, 139.
9. Markwardt, "Ancient Edessa," 29-32.
10. Anthony Ashley Beven, *The Hymn of the Soul Contained in the Syriac Acts of St. Thomas*, (Eugene, OR: ed. J. Armitage Robinson, Wipf & Stock, 2004), 25 – 27, verses 76 -78.
11. Ibid., 25- 27, verse 86.
12. *The New Interpreter's Study Bible.* 2237 (Rev. 17:14, 19:16)
13. Markwardt, "Ancient Edessa," 32.
14. Downey, *A History of Antioch,* 202-316.
15. Ibid., 312, 338-41.

H6
1. Markwardt, "Antioch and the Shroud," 5.
2. Markwardt, "Antioch and the Shroud," 6.
3. Alain Besancon (Translated by Jane Marie Todd), *The Forbidden Image* (Chicago, IL: The University of Chicago Press, 2000), 63-82.
4. Downey, *A History of Antioch,* 351.
5. "The First Council of Nicaea," *Catholic Encyclopedia,* http://www.newadvent.org/cathen/11044a.htm.
6. "First Council of Constantinople," *Catholic Encyclopedia,* http://www.newadvent.org/cathen/04308a.htm.
7. Downey, *A History of Antioch,* 352.
8. Ibid., 370.
9. Gustavus A. Eisen, *The Great Chalice of Antioch* (New York, NY: Kouchakji Freres, 1923), 5; Markwardt, "Antioch and the Shroud," 10.

H7
1. "Council of Chalcedon," *Catholic Encyclopedia,* http://www.newadvent.org/cathen/03555a.htm.
2. Downey, *A History of Antioch,* 456-475.
3. Heirich Pfeiffer, "The Shroud of Turin and the Face of Christ in Paleochristian, Byzantine and Western Medieval Art Part I," *Shroud Spectrum International, Issue 9, Part 4* (December 1983): 7 -21. http://www.shroud.com/pdfs/ssi09part4.pdf.
4. Downey, *A History of Antioch,* 519-525.
5. Ibid., 554
6. Ibid., 554.
7. John Moschos, *The Spiritual Meadow (Pratum Spirituale),*(Translated by John Eviratus, (Collegeville, MN: Liturgical Press, 2008), 212; Markwardt, "Modern Scholarship," 4.
8. Jack Markward, personal communication, October 8, 2016.

H8
1. Eisen, *The Great Chalice of Antioch*, 6; Markwardt, "Antioch and the Shroud," 12.
2. Glanville Downey, "Ephraemius, Patriarch of Antioch*," Church History, Vol. 7, No. 4. Published by Cambridge University Press on behalf of the American Society of Church History* (Dec. 1938): 364-370.
3. Markwardt, "Modern Scholarship and the History of the Turin Shroud," 19-20.
4. Downey, "Ephraemius, Patriarch of Antioch, 365.
5. Markwardt, "Modern Scholarship and the History of the Turin Shroud," 20.
6. Ibid., 20.
7. "Acheiropoieta," *Wikipedia,* http://en.wikipedia.org/wiki/Acheiropoieta.
8. Ernst Kitzinger, *The Cult of Images in the Age before Iconoclasm* (Dumbarton Oaks, Trustees for Harvard University, Dumbarton Oaks Papers, Vol. 8, 1954), 114,115.

H9
1. Markwardt, "Modern Scholarship and the History of the Turin Shroud," 20.
2. Ibid., 21.
3. Ibid., 21-23.
4. Ibid., 22.
5. Manolis Chatzidakis, "An Encaustic Icon of Christ at Sinai," *The Art Bulletin Volume XLIX Number Three* (September 1967) (Translated by Gerry Walters), 197-208.
6. Hans Belting, *Likeness and Presence: A History of the Image before the Era of Art*, Translated by Edmund Jephcotti, (Chicago and London: The University of Chicago Press, 1994), 133-34.
7. Alan D. Whanger and Mary Whanger, "Polarized image overlay technique: a new image comparison method and its application," *Applied Optics, Vol 24, No.6* (15 March 1985): 766-772.

	18.	Ibid., 244.
	19.	Ibid., 254.
	20.	Marzia Boi, "Palynology: Instrument of research for the relics of the Shroud of Turin and the Sudarium of Oviedo," Centro Internazionale di Sindonologiia (CIS), Turin, Italy, May 2, 2015, 1-16. http://www.shroud.com/pdfs/duemaggioBoiENG.pdf.
	21.	Emanuela Marinelli, "The question of pollen grains on the Shroud of Turin and the Sudarium of Oviedo," Valencia Shroud Conference, April 28-30, 2012, 1-13. https://www.shroud.com/pdfs/marinelli2veng.pdf.
L13	1.	Avinoam Danin, *Botany of the Shroud: The Story of Floral Images on the Shroud of Turin*, (Jerusalem, Israel: Danin Publishing (2010), 36-59.
	2.	Whanger, *Shroud of Turin*, 71-86.
	3.	Maloney, "A Contribution toward a History of Botanical Research on the Shroud of Turin," 264.
	4.	Whanger, *Shroud of Turin, 71.*
	5.	Danin, *Botany of the Shroud*, 8.
	6.	Ibid., 8.
	7.	Daniele Murra, and Paolo Di Lazzaro, "Sight and brain: an introduction to the visually misleading images," Proceedings of the International Workshop on the Scientific approach to the Acheiropoietos Images, ENEA Frascati, Italy, (4-6 May 2010): 4. http://www.acheiropoietos.info/proceedings/MurraWeb.pdf.
	8.	Daniele Murra, Paolo Di Lazzaro and Barrie Schwortz, "Perception of patterns after digital processing of low-contrast Images, the case of the Shroud of Turin," ENEA (2012): 1-17. http://opac.bologna.enea.it:8991/RT/2012/2012_12_ENEA.pdf.
L14	1.	Whanger, *Shroud of Turin*, 23-35.

Section 4: Image Characteristic Evidence

Intro	1.	Fanti, et al, "Evidences for Testing Hypotheses about the Body Image Formation of the Turin Shroud," 1-19.
	2.	Giulio Fanti, "Hypotheses Regarding the Formation of the Body Image on the Turin Shroud. A Critical Compendium," *Journal of Imaging Science and Technology* 55(6) 060507 (Nov.-Dec. 2011): 1-14.
C1	1.	Eric J. Jumper, et al, "A Comprehensive Examination of the Various Stains and Images on the Shroud of Turin," 451-53.
	2.	Schwalbe and Rogers, "Physics and Chemistry of the Shroud of Turin," 3-49.
C2	1.	Fanti, "Hypotheses Regarding the Formation of the Body Image," 3.
	2.	John P. Jackson, Eric J. Jumper, and William R. Ercoline, "Correlation of image intensity on the Turin Shroud with the 3-D structure of a human body shape," *Applied Optics Vol. 23, No. 14* (July 1984): 2265.
	3.	Evans, Mark, "1978 STURP Photomicrograph ME-29," *Shroud of Turin Website.*
	4.	Fanti, "Hypotheses Regarding the Formation of the Body Image," 3.
	5.	Jumper, et al, "A Comprehensive Examination", 450-53.
	6.	Rogers, *A Chemist's Perspective on the Shroud of Turin*, 44-45.
	7.	Fanti, et al., "Microscopic and Macroscopic Characteristics," 1-8.
	8.	Discussion with senior members of the TSC research team including John Jackson and Keith Propp. December 6, 2012.
C3	1.	Fanti, "Hypotheses Regarding the Formation of the Body Image," 3.
	2.	Nello Balossino, "Sul retro Della Sindone Non Vi E Impronta: Observazioni Directta Ed Elaborazioni Informatiche" *Sindon N.S. Quad. N. 19-20, (*December 2003), 57-69.
	3.	Paolo Di Lazzaro, Daniele Murra, Barrie Schwortz, "Pattern recognition after image processing of low-contrast images, the case of the Shroud of Turin, *ENEA Research Center of Frascati, Dept. Applications of Radiation, STERA Inc.,* (2012), 1-14.
	4.	Giuseppe Ghiberti, "Sindone le immagini 2002 Shroud Images," *ODPF, Torino, Italy* (2002): 1-4.
	5.	Giulio Fanti and Robert Maggiolo, "The Double Superficiality of the Frontal Image of the Turin Shroud," *Journal of Optics* Vol 6 No. 6 (2004): 491.
C4	1.	Giuseppe Ghiberti, "Sindone le immagini 2002 Shroud Images." 1-4.
	2.	Discussion at the TSC research center with senior members of the TSC research team including John Jackson and Keith Propp. December 6, 2012.
	3.	Fanti and Maggiolo, "The Double Superficiality of the Frontal Image of the Turin Shroud,"
C5	1.	Fanti, "Hypotheses Regarding the Formation of the Body Image," 3.

2. John H. Heller and Allan D. Adler, "A Chemical Investigation of the Shroud of Turin," *Canadian Society of Forensic Science Journal 14* (1981): 81-103, 95.

3. Jumper, et al, "A Comprehensive Examination," 456.

C6
1. Pellicori, S. and M.S. Evans, "The Shroud of Turin Through the Microscope," 34-43.
2. Mark Evans, "1978 STURP Photomicrograph ME-06," *Shroud of Turin Website*.
3. Fanti, "Hypotheses Regarding the Formation of the Body Image," 3.

B1
1. Fanti, "Hypotheses Regarding the Formation of the Body Image," 3.
2. Paul Vignon, *The Shroud of Christ,* 3-7.
3. Wilson, *The Shroud,* 17-21.

B2
1. Fanti, "Hypotheses Regarding the Formation of the Body Image," 3.
2. Jackson, Jumper, and Ercoline, "Correlation of image intensity on the Turin Shroud with the 3-D structure of a human body shape," 2246.
3. Ibid., 2246.

B3
1. Fanti, "Hypotheses Regarding the Formation of the Body Image," 3.
2. Jumper, et al, "A Comprehensive Examination," 450-53.
3. John P. Jackson, Eric J Jumper and W.R.Ercoline, "Three Dimensional Characteristics of the Shroud Image," *IEEE 1982 Proceedings of the International Conference on Cybernetics and Society* (October 1982): 559-575.
4. Jackson, Jumper and Ercoline, "Correlation of image intensity on the Turin Shroud with the 3-D structure of a human body shape," 2244-2269.
5. Jumper, et al, "A Comprehensive Examination," 451.

B4
1. Fanti, "Hypotheses Regarding the Formation of the Body Image," 3.
2. Giulio Fanti, Robert Basso, *The Turin Shroud: Optical Research in the Past, Present and Future,* (New York, NY: Nova Science Publishers, Inc., 2008), 5, 47, 63.
3. John P. Jackson, "The Vertical Alignment of the Frontal Image," *Shroud Spectrum International, No. 32/33* (Sept-Dec 1989): 1-23. (Updated 2014) http://www.shroudofturin.com/Resources/ShroudVerticalAlignmentoftheFrontalImage.pdf.
4. John P. Jackson, "Foreword to Vertical Alignment Paper," *Turin Shroud Center of Colorado Webpage* (Oct 2014): 1. http://shroudofturin.com/Resources/SDTV2.0ForwardtoVerticalAlignment.pdf

B5
1. Fanti et al, "Evidences for Testing Hypotheses about the Body Image Formation of the Turin Shroud," 9.
2. *The Shroud of Turin,* DVD, directed by David Rolfe (Performance Films for the BBC, London, UK., 2008). Discussion of side images, minute 14:51-15:20.

B6
1. Fanti, "Hypotheses Regarding the Formation of the Body Image," 3.
2. Jumper, et al, "A Comprehensive Examination," 450-53.
3. Heller and Adler, "Blood on the Shroud of Turin," 2742-2744.
4. David Ford, "The Shroud of Turin's 'Blood' Images: Blood, or Paint? A History of Science Inquiry," *Shroud of Turin Website* (2001): 1-11. http://www.shroud.com/pdfs/ford1.pdf.
5. Niels Svensson and Thibault Heimburger, "Forensic aspects and blood chemistry of the Turin Shroud Man," *Scientific Research and Essays Vol. 7(29)* (2012): 2513-2525. http://www.academicjournals.org/sre/PDF/pdf2012/30JulSpeIss/Svensson%20and%20Heimburger.pdf.
6. Kelly P. Kearse, "Blood on the Shroud of Turin: An Immunological Review," (2012): 1-22. http://www.shroud.com/pdfs/kearse.pdf.
7. Kelly P. Kearse, "A Critical Reevaluation of the Shroud of Turin Blood data: Strength of Evidence in the Characterization of the Bloodstains," *St. Louis, Mo. International Shroud Conference* (2014): 1-10. http://www.shroud.com/pdfs/stlkearsepaper.pdf.
8. Kelly Kearse and Thibault Heimburger, "The Shroud Blood Science of Dr. Pierluigi Baima Bollone: Another look at potassium, among other things," *Shroud of Turin Website* (21 January 2014): 1-10. https://shroudofturin.files.wordpress.com/2013/12/bbk-7.pdf.
9. Kelly P. Kearse, "Empirical evidence that the blood on the Shroud of Turin is of human origin: Is the current data sufficient?" *Shroud of Turin Website* (21 January 2013): 1-15. http://www.shroud.com/pdfs/kearse1.pdf.

B7
1. Fanti, "Hypotheses Regarding the Formation of the Body Image," 3.
2. Heller and Adler, "Blood on the Shroud of Turin," 19.
3. John P. Jackson, "Blood and Possible Images of Blood on the Shroud," *Shroud Spectrum International* (September 1987): 3-11. http://www.shroud.com/pdfs/ssi24part4.pdf.

B8	1.	Heller and Adler, " A Chemical Investigation of the Shroud of Turin," 81-103.
	2.	Jumper, et al, "A Comprehensive Examination", 460.
	3.	Fanti, "Hypotheses Regarding the Formation of the Body Image," 3.
B9	1.	Fanti, "Hypotheses Regarding the Formation of the Body Image," 3.
	2.	Zugibe, *The Crucifixion of Jesus, A Forensic Inquiry,* 129-163, 211-228.
	3.	Robert Bucklin, M.D., J.D., "The Legal and Medical Aspects of the Trial and Death of Jesus," *Medicine, Science and the Law Volume 10, No. 1* (January, 1970): 1-14. http://www.shroud.com/bucklin2.htm.
	4.	Bucklin, "An Autopsy," 1-4. http://www.shroud.com/bucklin.htm.
	5.	Robert Bucklin, "Postmortem Changes and the Shroud of Turin," *Shroud Spectrum International, #14* (March 1985): 1-6. http://www.shroud.com/pdfs/ssi14part3.pdf.
B10	1.	Fanti, "Hypotheses Regarding the Formation of the Body Image," 3.
	2.	Kevin Moran and Giulio Fanti, "Does the Shroud body image show any physical evidence of Resurrection?" *4th International Scientific Symposium, Centre International, Paris France,* (April 25-26 2002): 1-9. http://www.sindone.info/MORAN1.PDF.
B11	1.	Giles F. Carter, "Formation of the Image on the Shroud of Turin by x-rays: a New Hypotheses," in *Archaeological Chemistry, The American Chemical Society* 0065-2393/84/0205-0425806-50.0 (1984): 431-33.
	2.	August D. Accetta, Kenneth Lyons and John P. Jackson, "Nuclear Medicine and its Relevance to the Shroud of Turin," *World Wide Congress on the Shroud of Turin, Oviedo, Spain,* (2000/2002): 1-7. https://shroud.com/pdfs/accett2.pdf.
	3.	Discussion with senior members of the TSC research team including John Jackson and Keith Propp. December 6, 2012.
	4.	John P. Jackson, Presentation on Shroud to Roman Catholic Forum, WCBOhio (2002): http://www.youtube.com/watch?v=8iOlCzBdrhc. http://www.youtube.com/watch?v=2rYr774TUI8.
	5.	Whanger and Whanger, "Polarized image overlay technique: a new image comparison method and its application," 766-772.
	6.	Whanger and Whanger, *The Shroud of Turin: An Adventure in Discovery,* 111-115.
	7.	Ibid., 115-117.

Section 5: Image-Formation Hypotheses

F1	1.	Fanti, "Hypotheses Regarding the Formation of the Body Image," 1-3.
	2.	Vignon, *The Shroud of Christ*, 72-91.
	3.	Jackson, Jumper and Ercoline, "Correlation of image intensity on the Turin Shroud with the 3-D structure of a human body shape," 2253-63.
	4.	Paul Vignon, "The Problem of the Holy Shroud," *Scientific American* (March 1, 1937): 162.
F2	1.	Vignon, *The Shroud of Christ*, 93-101.
	2.	Rogers, *Chemist's Perspective, 28-35, 38, 99-120.*
	3.	Raymond N. Rogers and Anna Arnoldi, "The Shroud of Turin: An Amino-Carbonyl Reaction (Maillard Reaction) May Explain the Image Formation," This article originally appeared in *Melanoidins vol. 4, Office for Official Publications of the European Communities, Luxembourg* (2003): 1-9. http://www.shroud.com/pdfs/rogers7.pdf.
	4.	Jackson, Jumper and Ercoline, "Correlation of image intensity on the Turin Shroud with the 3-D structure of a human body shape," 2263-64.
	5.	Heller and Adler, "A Chemical Investigation of the Shroud of Turin," 81-103.
	6.	Giulio Fanti, "Comments on gas diffusion hypothesis," *University of Padua, Italy* (2004); 1-10. https://www.dii.unipd.it/~giulio.fanti/research/Sindone/diffusion.pdf.
	7.	Vignon, "The Problem of the Holy Shroud," 162.
	8.	Ibid., 163.
	9.	Heller and Adler, "A Chemical Investigation," 81-103.
F3	1.	Jackson, Jumper and Ercoline, "Correlation of image intensity on the Turin Shroud with the 3-D structure of a human body shape," 2251-53.
	2.	Walter C. McCrone, *Judgment Day for the Shroud of Turin* (Amherst, NY: Prometheus Books, 1999).
	3.	Giulio Fanti, Personal communication, 12/02/2016 and 8/08/2016.
	4.	Walter C. McCrone, "The Shroud of Turin: Blood or Artist's Pigment?", Accounts of Chemical Research, American Chemical Society, 23 (1990); pp 77-83.

REFERENCES

5. R.A. Morris, L.A. Schwalbe and J.R. London, "X-ray Fluorescence investigation of the Shroud of Turin," *X-Ray Spectrometry Volume 9, Issue 2* (April 1980): 40-47.

6. McCrone, *Judgment Day for the Shroud of Turin.*.

F4
1. Emily A Craig, PhD, and Randall R. Bresee, PhD, "Image Formation and the Shroud of Turin," *Reprinted from the Journal of Imaging Science and Technology* (February 1994): 1-18. http://www.shroud.com/pdfs/craig.pdf.
2. Fanti, "Hypotheses Regarding the Formation of the Body Image," 4.
3. Jackson, Jumper and Ercoline, "Correlation of image intensity on the Turin Shroud with the 3-D structure of a human body shape," 2251-53.
4. Isabel Piczek Special Comments, "Four New Theories of How the Shroud Image May have Been Formed," *British Society for the Turin Shroud-BSTS* (March/April 1994) 5. http://www.shroud.com/pdfs/n37part3.pdf.

F5
1. Fanti, "Hypotheses Regarding the Formation of the Body Image," 4.
2. Jackson, Jumper and Ercoline, "Correlation of image intensity on the Turin Shroud with the 3-D structure of a human body shape," 2251-53, 2265-2267.
3. Fanti and Basso, *The Shroud: Optical Research,* 51-52.
4. *The Shroud of Turin,* DVD, Interview with John Jackson, concerning Jackson experiments with bas-relief method, minute17:30-19:06 .

F6
1. Luigi Garlaschelli, "Life-size Reproduction of the Shroud of Turin and its Image," *Journal of Imaging Science and Technology* (2010), 1-11.
2. Giulio Fanti and Thibault Heimburger, "Letter to the editor, Comments on 'Life-size Reproduction of the Shroud of Turin and its image' by L. Garlaschelli," *Journal of Imaging Science and Technology* (Mar-Apr. 2011): 1-3. http://www.ingentaconnect.com/content/ist/jist/2011/00000055/00000002/art00002.
3. Fanti, "Hypotheses Regarding the Formation of the Body Image," 4.
4. Thibault Heimburger, "Comments About the Recent Experiment of Professor Luigi Garlaschelli," (2009), 1-9. *Shroud of Turin Website* (2009): 1-9. . http://www.shroud.com/pdfs/thibault-lg.pdf.
5. Thibault Heimburger and Giulio Fanti, "Scientific Comparison Between the Turin Shroud and the First hand made Whole Copy," *Proceedings of the International Workshop on the Scientific Approach to the Acheiropoietos Image, ENEA Frascati, Italy* (4-6 May 2010): 1-10. http://www.acheiropoietos.info/proceedings/HeimburgerWeb.pdf.
6. Jackson, Jumper and Ercolinel, "Correlation of image intensity on the Turin Shroud with the 3-D structure of a human body shape," 2251-53.
7. Reuters News Service On-line, "Italian Scientist reproduces Shroud of Turin". http://www.reuters.com/article/us-italy-shroud-idUSTRE5943HL20091005.
8. Catholic News Agency On-line, " Experts question scientists claim of reproducing Shroud of Turin." http://www.catholicnewsagency.com/news/experts_question_scientists_claim_of_reproducing_shroud_of_turin/.
9. Garlaschelli, "Life-size Reproduction of the Shroud of Turin and its Image," 8.

F7
1. Nicholas P. L. Allen, "Verification of the Nature and Causes of the Photo-negative Images on the Shroud of Lirey-Chambéry-Turin," *De Arte 51, Pretoria, UNISA* (1995): 21-35. http://www.sunstar-solutions.com/AOP/esoteric/Images_on_the_Shroud_of_Turin.htm.
2. Estelle A Maré, "Science, Art History and the Shroud of Turin: Nicolas Allen's research on the iconography and production of the image of a crucified man," *South African Journal of Art History Volume 14* (1999): 66-83.
3. Barrie M. Schwortz, "Is the Shroud of Turin a Medieval Photograph? A Critical Examination of the Theory," *The Worldwide Shroud Conference Ovieto Spain* (27,28,29 Aug 2000): 1-10. http://www.shroud.com/pdfs/orvieto.pdf.
4. Wilson, *The Blood,* 209, 210-218; fig 25, pl. 47.
5. Oxley, *The Challenge of the Shroud*, 248,-250.
6. Alan D. Adler, "The Nature of the Body Images on the Shroud of Turin", *Shroud of Turin Website* (1999): 6. http://www.shroud.com/pdfs/adler.pdf.
7. Allen, "Verification of the Nature and Causes," 21 (page 1 in reprint.)
8. Schwortz, "Is the Shroud of Turin a Medieval Photograph?" 2,

F8
1. Nathan Wilson website on Shadow Hypothesis: http://www.shadowshroud.com/.
2. Ibid.

F9
1. John P. Jackson, "Is the Image on the Shroud Due to a Process Heretofore Unknown to Modern Science?" *the International Scientific Symposium, Paris, France* (1989): 1-29. http://shroudofturin.com/Resources/ShroudFallThroughSDTV2.0.pdf
2. John P. Jackson, "Forward to Fall-Through Hypothesis," *Shroud Center of Colorado Website* (2014): 1-2. http://www.shroudofturin.com/Resources/ShroudForwardtoFallThrough.pdf.
3. Antonacci, *Test the Shroud at the Atomic and Molecular Levels,* 233-276.

4. Ibid., 249.

F10
1. Giulio Fanti, "Body Image-Formation Hypotheses Based on Corona Discharge: Discussion," International Conference on the Shroud of Turin "Perspectives of a Multifaceted Enigma", Columbus, Ohio (2008): 1-28. http://ohioshroudconference.com/papers/p15.pdf."
2. Giulio Fanti, "Can a Corona Discharge Explain the Body Image of the Turin Shroud?." Journal of Imaging Science and Technology, Mar-Apr 2010; 1-11.
3. Adler, "The Nature of the Body Images on the Shroud of Turin", 8. http://www.shroud.com/pdfs/adler.pdf
4. Giulio Fanti, Francesco Lattarulo, Giancarlo Pesavento, "Experimental results using corona discharge to attempt to reproduce the Turin Shroud Image", *Bari Shroud Conference* (2014): 1-14.
5. Paolo Di Lazzaro, Daniele Murra, Enrico Nichelatti, Antonino Santoni, and Giuseppe Baldacchini, "Superficial and Shroud-like colorization of linen by short laser pulses in the vacuum ultraviolet", in *Applied Optics, Optical Society of America* Vol. 51, No. 36 (20 December 2012): 8567-8578.
6. D.S. Spicer, E.T. Toton, "Electric Charge Separation as the Mechanism for Image Formation on the Shroud of Turin: A Natural Mechanism", *St. Louis Shroud Conference*, (Revised May 2015): 1-18. http://www.shroud.com/pdfs/stlspicerpaper.pdf

Section 7: Dating the Shroud

1. STURP, *Formal Proposal for Performing Scientific Research on the Shroud of Turin*, (Shroud of Turin Research Project, Inc., (1984): 1-177.
2. Harry E. Gove, *Relic, Icon or Hoax?: Carbon Dating the Turin Shroud* (Philadelphia, PA: Institute of Physics Publishing,1996), 6-7.
3. William Meacham, "Radiocarbon Measurement and the Age of the Turin Shroud: Possibilities and Uncertainties," *Proceedings of the Symposium "Turin Shroud – Image of Christ?" Hong Kong* (March 1986): 1-16. http://www.shroud.com/meacham.htm.
4. Mark Antonacci, *The Resurrection of the Shroud* (New York, NY: M. Evans and Company, Inc., 2000), 157.
5. Wilson, *The Shroud*, 97.
6. Meacham, "Radiocarbon Measurement," 3.
7. Gove, *Relic, Icon or Hoax*, 194.
8. Ibid., 113, 165.
9. Ibid., 174-175.
10. Ibid., 201.
11. Ibid., 219.
12. Antonacci, *The Resurrection of the Shroud*, 181.
13. Wilson, *The Shroud*, 87.
14. P. E. Damon, et al, "Radiocarbon Dating of the Shroud of Turin", *Nature* 337 (16 February 1989): 611-615.
15. Wilson, *The Shroud*, 89.
16. Damon, et al, "Radiocarbon Dating of the Shroud of Turin," 611-615.
17. Wilson, *The Shroud*, 2, 83.
18. Rachel A. Freer-Waters and A.J. Timothy Jull, "Investigating a Dated Piece of the Shroud of Turin," *Radiocarbon, Vol. 52, No 4*. (2010): 1521-1527. https://journals.uair.arizona.edu/index.php/radiocarbon/article/view/3419.
19. Mark Oxley, "Evidence is not Proof: A Response to Prof. Timothy Jull," *Shroud of Turin Website* (2011), 9. http://www.shroud.com/pdfs/oxley.pdf.
20. John P. Jackson, personal communication based on data recorded in STURP experimental result notebook (Aug 15, 2015).
21. Bryan J. Walsh, "The 1988 Shroud Radiocarbon Tests Reconsidered (Parts 1 and 2),"*The Proceedings of the 1999 Shroud of Turin International Research Conference, Richmond, Virginia* (1999).
22. Marco Riani, Anthony C. Atkinson, Giulio Fanti, Fabio Crosilla, "Regression analysis with partially labeled regressors: carbon dating of the Shroud of Turin," *Statistical Computing No. 23* (2013): 551-561.
23. Fanti and Malfi, *The Shroud of Turin: First Century after Christ*, 209-246.
24. John P. Jackson, et al, "On the scientific validity of the Shroud's radiocarbon date," *The Proceedings of the 1999 Shroud of Turin International Research Conference, Richmond, Virginia* (1999).
25. John P. Jackson, et al, "On Middle Byzantine History of the Turin Shroud," *The Proceedings of the 1999 Shroud of Turin International Research Conference, Richmond, Virginia* (1999).
26. Alan Adler, "Further Spectroscopic Investigation of the Samples from the Shroud of Turin," *Taped Presentation at Shroud Conference in Turin, Italy* (1998): 2:00-4:15; 17:40-18:08; 22:40-23:17. http://shrouduniversity.com/libraryaudio.php.
27. Rogers, *A Chemists Perspective on the Shroud of Turin*: 26.

28. Ray Rogers, "Studies on the Radiocarbon Samples from the Shroud of Turin," *Thermochimica Acta 425* (2005): 189, 191, 192, 193.
29. M. Sue Benford and Joseph G. Marino, "Historical Support of a 16th Century Restoration in the Shroud C-14 Sample Area," *Shroud of Turin Website* (2002): 1-8. http://www.shroud.com/pdfs/histsupt.pdf.
30. M. Sue Benford and Joseph Marino, "New Historical Evidence Explaining the "Invisible Patch" in the 1988 C-14 Sample Area of the Turin Shroud," *Shroud of Turin Website* (2005): 1-16. http://www.shroud.com/pdfs/benfordmarino.pdf.
31. Leoncia A. Garza-Valdes, *The DNA of God?*, (New York, NY: Doubleday, 1999), 47-54.
32. Jackson, et al, "On the scientific validity of the Shroud's radiocarbon date," 14.
33. John P. Jackson, "A New Radiocarbon Hypothesis," *Shroud of Turin Website* (2008): 1-2. https://www.shroud.com/pdfs/jackson.pdf
34. *The Shroud of Turin,* DVD, Interview with Christopher Ramsey at 57:19 – 57:54.
35. Thomas J. Phillips, "Shroud irradiated with neutrons?" *Nature 337* (16 February 1989): Letter to the Editor, 594.
36. R. E. M. Hedges, "Shroud irradiated with neutrons? Response," *Nature 337* (16 February 1989): Letter to the Editor, 594.
37. Phillips, "Response to Hedges," (The Letter that '*Nature*' did not print) *Shroud of Turin Website.* http://www.shroud.com/pdfs/n22part5.pdf

Conclusion

1. The Light of the World Painting, Wikipedia, the free encyclopedia. https://en.wikipedia.org/wiki/The_Light_of_the_World_(painting)

Appendix 2: Rating Details for Image-Formation Hypotheses

1. Heller and Adler, "A Chemical Investigation of the Shroud of Turin,"
2. Rogers, *A Chemist's Perspective on the Shroud of Turin,* 115.
3. Jackson, Jumper and Ercoline, "Correlation of image intensity on the Turin Shroud with the 3-D structure of a human body shape," 2264.
4. Ibid., 2252.
5. Isabel Piczek, "Noted Los Angeles artist Isabel Piczek replies to the Craig Bresee Theory," *BSTS Newsletter No. 37 – Part 3* (March – April, 1994): 5. http://www.shroud.com/pdfs/n37part3.pdf.
6. *The Shroud of Turin,* DVD, Interview with John Jackson at 18:30 - 19:10.
7. Ibid., 19:10 - 20:00.
8. Fanti and Basso, *The Turin Shroud: Optical Research*, 50 -52.
9. Fanti and Heimburger, "Letter to the editor, Comments on 'Life-size Reproduction of the Shroud of Turin and its image' by L. Garlaschelli," 2.
10. Heimburger, "Comments About the Recent Experiment of Professor Luigi Garlaschelli," 4.
11. Jackson, "Is the Image on the Shroud Due to a Process heretofore Unknown to Modern Science?"
12. Ibid., 10.
13. Ibid., 14.
14. Jackson, Jumper and Ercoline, "Correlation of image intensity on the Turin Shroud with the 3-D structure of a human body shape," 2264-2265.
15. Fanti, "Can a Corona Discharge Explain the Body Image of the Turin Shroud?." 1-11.
16. Ibid., 5.

Photograph and Image Credits

(1) By Anonymous, [Public domain], Creative Commons (Wikimedia) (PD-Art).

(2) ©Barrie Schwortz Collection, STERA, Inc.

(3) ©Barrie Schwortz Collection, STERA, Inc.

(4) ©TSC Image Collection.

(5) ©TSC Image Collection.

(6) ©By Giovanni Dall'Orto (Own work) [Attribution], Creative Commons (Wikimedia). This file has been identified as being free of known restrictions under copyright law, including all related and neighboring rights.

(7) © Aldo Guerreschi. Used with permission.

(8) Sarcophagus #151 Lateran Collection. Source Heinrich Pfeiffer, S.J. article *"The Shroud of Turin and the Face of Christ in Paleochristian, Byzantine and Western Art"*, Shroud Spectrum International Dec 1983, Issue 9. This file has been identified as being free of known restrictions under copyright law, including all related and neighboring rights.

(9) ©Barrie Schwortz Collection, STERA, Inc.

(10) By Anonymous, [Public domain], Creative Commons (Wikimedia) (PD-ART).

(11) By Anonymous, [Public domain], Creative Commons (Wikimedia) (PD-ART).

(12) By Anonymous, [Public domain], Creative Commons (Wikimedia) (PD-ART).

(13) By Meister des Rabula-Evangeliums [Public domain], The work of art depicted in this image and the reproduction thereof are in the public domain worldwide. The reproduction is part of a collection of reproductions compiled by The Yorck Project. The compilation copyright is held by Zenodot Verlagsgesellschaft mbH and licensed under the GNU Free document License. (PD-ART).

(14) Coin image courtesy of Giulio Fanti. Shroud face image ©Barrie Schwortz Collection, STERA, Inc.

(15) By Anonymous, [Public domain], Creative Commons (Wikimedia) (PD-ART).

(16) ©Centro Espanol de Sindonologia (CES) Spain, furnished courtesy of Barrie Schwortz, STERA, Inc.

(17) The work of art depicted in this image and the reproduction thereof are in the public domain worldwide. The reproduction is part of a collection of reproductions compiled by The Yorck Project. The compilation copyright is held by Zenodot Verlagsgesellschaft mbH and licensed under the GNU Free document License. (PD-ART).

(18) ©RassaphoreGeorge, Creative Commons: This file is made available under the Creative Commons CC0.10 Universal Public Domain Dedication.

(19) By Anonymous, [Public domain], Creative Commons (Wikimedia) (PD-ART).

(20) By Anonymous, [Public domain], Creative Commons (Wikimedia).(PD-ART).

(21) ©Barrie Schwortz Collection, STERA, Inc.

(22) By Anonymous, [Public domain], Creative Commons (Wikimedia) (PD-ART).

(23) By Naddo Ceccarelli, ca. 1347. [Public domain], Creative Commons (Wikimedia) (PD-ART).

(24) ©1998 Ian Wilson, Used with permission.

(25) ©Barrie Schwortz Collection, STERA, Inc.

(26) Giovanni Bellini's Pietà [Public domain], Creative Commons (Wikimedia) (PD-ART).

(27) Byzantine Icon before 1499. [Public domain], Creative Commons (Wikimedia) (PD-ART).

(28) ©Stanislav Trykov. This file is licensed under the Creative Commons Attribution-Share Alike 3.0 Unported license.

(29) By Anonymous, [Public domain], Creative Commons (Wikimedia) (PD-ART).

PHOTOGRAPH AND IMAGE CREDITS

(30) ©B. Didier (own work), via Wikimedia Commons. This file is licensed under the Creative Commons Attribution-Share Alike 2.5 Generic license.

(31) ©Niels Svensson, used with permission.

(32) ©Ian Wilson, used with permission.

(33) ©Moi-même. Photograph released to the public domain, Creative Commons (Wikimedia).

(34) Photograph by unknown photographer. (Public Domain), Creative Commons (Wikimedia).

(35) ©Barrie Schwortz Collection, STERA, Inc.

(36) Photograph by Secundo Pia. Public Domain.

(37) ©Rubén Betanzo S. released to the public domain. Creative Commons (Wikimedia).

(38) ©Barrie Schwortz Collection, STERA, Inc.

(39) ©U.S. Air Force Photo/Mike Kaplan, Released to the Public Domain.

(40) ©Barrie Schwortz Collection, STERA, Inc.

(41) ©Barrie Schwortz Collection, STERA, Inc.

(42) ©Westmere32nm (Public Domain), Creative Commons (Wikimedia).

(43) ©TSC Image Collection.

(44) Courtesy of Sindone.org photo gallery, official site of the Holy Shroud created and sponsored by the Archdiocese of Turin, Italy.

(45) ©Barrie Schwortz Collection, STERA, Inc.

(46) ©Barrie Schwortz Collection, STERA, Inc.

(47) ©Barrie Schwortz Collection, STERA, Inc.

(48) ©Barrie Schwortz Collection, STERA, Inc.

(49) Reconstruction of Roman Flagrum from p. 56, fig. 27 in book by Paul Vignon: *Le Saint Suaire de Turin devant la Science, l'Archéologie, l'Histoire, l'Iconographie, la Logique,* Paris, France, Masson, 1939.

(50) ©Barrie Schwortz Collection, STERA, Inc.

(51) ©Barrie Schwortz Collection, STERA, Inc.

(52) ©Barrie Schwortz Collection, STERA, Inc.

(53) ©Barrie Schwortz Collection, STERA, Inc.

(54) ©Barrie Schwortz Collection, STERA, Inc.

(55) ©Takkk. This file is licensed under the Creative Commons Attribution-Share Alike 3.0 Unported license.

(56) By Anonymous, [Public domain], Creative Commons (Wikimedia) (PD-ART). Mosaics in Hosios Loukas Monastery, Boeotia, Greece.

(57) ©Barrie Schwortz Collection, STERA, Inc.

(58) ©Barrie Schwortz Collection, STERA, Inc.

(59) ©Barrie Schwortz Collection, STERA, Inc.

(60) ©Barrie Schwortz Collection, STERA, Inc.

(61) ©Barrie Schwortz Collection, STERA, Inc.

(62) ©Barrie Schwortz Collection, STERA, Inc.

(63) ©Barrie Schwortz Collection, STERA, Inc.

PHOTOGRAPH AND IMAGE CREDITS

(64) ©Barrie Schwortz Collection, STERA, Inc.
(65) ©Barrie Schwortz Collection, STERA, Inc
(66) Courtesy of David Rolfe, Performance Films.
(67) ©1998 Ian Wilson, Used with permission.
(68) Courtesy of David Rolfe, Performance Films.
(69) ©Barrie Schwortz Collection, STERA, Inc.
(70) ©Barrie Schwortz Collection, STERA, Inc.
(71) ©Barrie Schwortz Collection, STERA, Inc.
(72) ©Barrie Schwortz Collection, STERA, Inc.
(73) ©Aldo Guerreschi. Used with permission.
(74) ©Aldo Guerreschi. Used with permission.
(75) This work has been released into the public domain by its author, Dartmouth College Electron Microscope Facility.
(76) ©Barrie Schwortz Collection, STERA, Inc.
(77) Courtesy of Giulio Fanti and the University of Padua Research Project: Padua, Italy 2008 #CPDA099244
(78) ©Mariana Ruiz. This work has been released into the public domain by its author.
(79) ©Barrie Schwortz Collection, STERA, Inc.
(80) ©Barrie Schwortz Collection, STERA, Inc.
(81) ©Barrie Schwortz Collection, STERA, Inc.
(82) ©Barrie Schwortz Collection, STERA, Inc.
(83) ©Barrie Schwortz Collection, STERA, Inc.
(84) ©Barrie Schwortz Collection, STERA, Inc.
(85) ©TSC Image Collection.
(86) Courtesy of David Rolfe, Performance Films.
(87) ©Barrie Schwortz Collection, STERA, Inc.
(88) ©Barrie Schwortz Collection, STERA, Inc.
(89) ©Barrie Schwortz Collection, STERA, Inc.
(90) ©1998 Mary and Alan Whanger.
(91) ©Barrie Schwortz Collection, STERA, Inc.
(92) ©Barrie Schwortz Collection, STERA, Inc.
(93) ©Barrie Schwortz Collection, STERA, Inc.
(94) ©Barrie Schwortz Collection, STERA, Inc.
(95) ©Sue Benford and Joe Marino, left side of figure / ©Barrie Schwortz Collection, STERA, INC, right side of figure
(96) Courtesy of Ian Wilson (original publication Telegraph Group Ltd, London, 1988).
(97) Courtesy of David Rolfe, Performance Films.
(98) (Public domain), Creative Commons (Wikimedia) (PD-ART).
(99) ©TSC Image Collection.

PHOTOGRAPH AND IMAGE CREDITS

(100) ©TSC Image Collection.

(101) ©TSC Image Collection.

(102) ©TSC Image Collection.

(103) ©TSC Image Collection

Every effort has been made to obtain permissions for various illustrations and photographs used in this document at the time of publishing, but in the case of any omissions or oversights we will be pleased to make the appropriate acknowledgments in any subsequent version or edition.

TSC wishes to especially acknowledge the generous help of several individuals who have made major contributions to Shroud research and who have helped us by providing images for this publication. First, we would like to thank the well-known Shroud historian Ian Wilson for providing four important images, the image of the Jackson Shroud lifting device (images 24 and 67), the Templecombe panel painting (image 32) and the photograph from the carbon dating news conference (image 96). Next, we would like to thank the Dutch physician and Shroud author Niels Svensson for generously providing the high definition image of the Lirey Pilgrim Medallion (image 31). David Rolfe who heads Performance Films generously made available four high definition images (images 66, 68, 86, 97) from his outstanding Shroud films. Professional photographer and Shroud Image expert Aldo Guerreschi generously provide three images related to the ancient water stains on the linen fabric of the Shroud (images 7, 73, 74). We want to especially thank Joe Marino and his late beloved wife, Sue Benford, for the image (95) that illustrates where the Raes sample and the adjacent radio carbon testing samples were cut from the Shroud. The respected and well-known Italian research scientist, Giulio Fanti, has offered many valuable comments and suggestions and for this version of the Critical Summary and he has graciously provided two important high definition images as well. The first image (14) is a high-resolution image of the Jesus image-bearing solidus coin of Justinian II dated from 692. AD. The second image is equally important. It is a scanning electron microscope photograph of a pollen grain that was vacuumed from between the Shroud and its backing cloth. Finally, we would like to extend our special thanks to Barry Schwortz and his STERA organization. Barry was a colleague of John Jackson on the STURP project and was responsible for documenting photography. He did a wonderful job of that, and today runs the important Shroud research repository website Shroud.com. We thank Barrie for providing the lion share of the high definition Shroud images used throughout this edition of the Critical Summary

We also wish to honor the memory of Jim Gilvoy of CMJ Marian Publishers. Jim provided invaluable guidance along the way as we worked toward the publication of this work. Jim is gone but not forgotten. He will always be in our hearts.